Practical Machine Learning

The book provides an accessible, comprehensive introduction for beginners to machine learning, equipping them with the fundamental skills and techniques essential for this field.

It enables beginners to construct practical, real-world solutions powered by machine learning across diverse application domains. It demonstrates the fundamental techniques involved in data collection, integration, cleansing, transformation, development, and deployment of machine learning models. This book emphasizes the importance of integrating responsible and explainable AI into machine learning models, ensuring these principles are prioritized rather than treated as an afterthought. To support learning, this book also offers information on accessing additional machine learning resources such as datasets, libraries, pre-trained models, and tools for tracking machine learning models.

This is a core resource for students and instructors of machine learning and data science looking for a beginner-friendly material which offers real-world applications and takes ethical discussions into account.

Practical Machine Learning
A Beginner's Guide with Ethical Insights

Ally S. Nyamawe, Mohamedi M. Mjahidi,
Noe E. Nnko, Salim A. Diwani, Godbless G. Minja,
and Kulwa Malyango

CRC Press is an imprint of the
Taylor & Francis Group, an **informa** business

A CHAPMAN & HALL BOOK

Designed cover image: KHAIDARY HADAIKA

First edition published 2025
by CRC Press
2385 NW Executive Center Drive, Suite 320, Boca Raton FL 33431

and by CRC Press
4 Park Square, Milton Park, Abingdon, Oxon, OX14 4RN

CRC Press is an imprint of Taylor & Francis Group, LLC

© 2025 Ally S. Nyamawe, Mohamedi M. Mjahidi, Noe E. Nnko, Salim A. Diwani, Godbless G. Minja, and Kulwa Malyango

Reasonable efforts have been made to publish reliable data and information, but the author and publisher cannot assume responsibility for the validity of all materials or the consequences of their use. The authors and publishers have attempted to trace the copyright holders of all material reproduced in this publication and apologize to copyright holders if permission to publish in this form has not been obtained. If any copyright material has not been acknowledged please write and let us know so we may rectify in any future reprint.

The Open Access version of this book, available at www.taylorfrancis.com, has been made available under a Creative Commons [Attribution-Non Commercial-No Derivatives (CC-BY-NC-ND)] 4.0 license.

Any third party material in this book is not included in the OA Creative Commons license, unless indicated otherwise in a credit line to the material. Please direct any permissions enquiries to the original rightsholder.

This work was carried out with the aid of a grant from the Artificial Intelligence for Development in Africa Program, a program funded by Canada's the International Development Research Centre, Ottawa, Canada and the Swedish International Development Cooperation Agency.

Trademark notice: Product or corporate names may be trademarks or registered trademarks and are used only for identification and explanation without intent to infringe.

Library of Congress Cataloging-in-Publication Data
Names: Nyamawe, Ally S., author. | Mjahidi, Mohamedi M., author. | Nnko, Noe E. (Noe Elisa), author. | Diwani, Salim A., author. | Minja, Godbless G., author. | Malyango, Kulwa, author.
Title: Practical machine learning : a beginner's guide with ethical insights / Ally S. Nyamawe, Mohamedi M. Mjahidi, Noe E. Nnko, Salim A. Diwani, Godbless G. Minja, Kulwa Malyango.
Description: First edition. | Boca Raton, FL : CRC Press, 2025. | Includes bibliographical references and index.
Identifiers: LCCN 2024034546 (print) | LCCN 2024034547 (ebook) | ISBN 9781032782164 (hardback) | ISBN 9781032770291 (paperback) | ISBN 9781003486817 (ebook)
Subjects: LCSH: Machine learning. | Artificial intelligence.
Classification: LCC Q325.5 .N93 2025 (print) | LCC Q325.5 (ebook) | DDC 006.3/1--dc23/eng/20241119
LC record available at https://lccn.loc.gov/2024034546
LC ebook record available at https://lccn.loc.gov/2024034547

ISBN: 978-1-032-78216-4 (hbk)
ISBN: 978-1-032-77029-1 (pbk)
ISBN: 978-1-003-48681-7 (ebk)

DOI: 10.1201/9781003486817

Typeset in Times
by SPi Technologies India Pvt Ltd (Straive)

Contents

About the authors vi
Preface ix
Acknowledgments x
Glossary xi

1	Fundamentals of machine learning	1
2	Mathematics for machine learning	18
3	Data preparation	76
4	Machine learning operations	95
5	Machine learning software and hardware requirements	110
6	Responsible AI and explainable AI	138
7	Artificial general intelligence	152
8	Machine learning step-by-step practical examples	162

Appendix: Machine learning resources 207
Index 210

About the authors

Ally S. Nyamawe is a Computer Scientist with over 15 years of experience in academia. He holds a PhD in Computer Science and Technology from Beijing Institute of Technology (2020), and his research interest mainly focuses on AI applications in Software Engineering. Nyamawe is a Senior Lecturer in Computer Science at the University of Dodoma, Tanzania. Nyamawe's recent work focused on contributing to developing AI-driven innovations that address social challenges and AI uptake for sustainable development in Africa. Nyamawe has been working on different research projects committed to fostering the application of AI for social good and leveraging coding and algorithmic skills in addressing real-world problems. Nyamawe has extensive experience in leading projects with support from the World's renowned funders, including IDRC, Sida, UNESCO-TWAS, and the EU Erasmus+ Program. Nyamawe actively contributes to the academic community through publications and participation in renowned conferences and international forums. He has served on the program committees for prestigious conferences, including the 37th IEEE/ACM International Conference on Automated Software Engineering, the 11th International Workshop on Software and Systems Traceability, and the 1st International Conference on the Advancements of Artificial Intelligence in African Context (AAIAC 2023). His recent recognition includes a 2022 recipient of the Seed Grant for New African Principal Investigators awarded by The World Academy of Sciences under UNESCO funding.

Mohamedi M. Mjahidi (PhD) is a Lecturer at the Department of Computer Science and Engineering (DoCSE), College of Informatics and Virtual Education (CIVE), the University of Dodoma (UDOM), Dodoma, Tanzania. He graduated from the University of Dar es Salaam (UDSM) in 2006 with a BSc in Computer Science and completed his MSc in Telecommunication Engineering at UDOM in 2011. He then completed his PhD in Computer Engineering at the Gazi University, Ankara, Turkey, in 2020. His research interests include Artificial Intelligence, Machine Learning, and Computer and Mobile Networks. At the time of writing this book, Mjahidi is serving as the Lab and Training Coordinator for AI4D Research Lab.

About the authors

Noe E. Nnko currently serves as the Acting Director of ICT at the University of Dodoma. He is an experienced Cybersecurity, Artificial Intelligence (AI), and Data Science Researcher/Engineer, boasting over 11 years of experience in telecommunication, networking, software, and web application security. He specializes in the design and implementation of AI models, particularly for detecting anomalies in computer networks. Additionally, he has expertise in using blockchain technology to develop secure and privacy-preserving decentralized systems. His PhD research at Northumbria University, UK, focused on exploiting blockchain technology and Artificial Immune Systems (AIS) to create a decentralized, secure, and privacy-preserving e-Government system for enforcing data protection and trust. During his master's degree studies in India in 2014, he received professional training in ethical hacking, Linux server administration, Android mobile application development, CCNA, and web application programming in JAVA. One of his notable contributions to AI and academic research is the development of a new general-purpose multiclass classifier based on the Dendritic Cell Algorithm (McDCA), which is currently under review by the *IEEE Transactions on Neural Networks and Learning Systems* journal. His current research and practical endeavors focus on leveraging African-origin datasets to develop AI-powered solutions that address privacy concerns related to data breaches and unlawful access to personal information.

Salim A. Diwani is a highly skilled and experienced Lecturer at the University of Dodoma, specializing in Machine Learning and Artificial Intelligence for more than 12 years. Diwani, acknowledged as a senior expert in various fields, has made noteworthy contributions to both academia and practical applications. At the University of Dodoma, he has fostered a vibrant atmosphere in which students have developed and put into practice cutting-edge artificial intelligence solutions across various industries, including agriculture and healthcare. Diwani not only holds a position at the institution but also acts as the coordinator for the Healthcare Coordination Unit at AI4D Research Lab. He is responsible for supervising a group of committed professionals who are dedicated to utilizing AI technologies to tackle urgent healthcare issues. Diwani's leadership includes the responsibility of hosting graduate students supported by AI4D Research Lab. Diwani is leading the Healthcare Coordination Unit in developing AI solutions specifically designed for the local requirements of the Anglophone region in Africa. Diwani and his colleagues are closely collaborating with the Ministry of Health in the Government of Tanzania to build an AI policy in the health sector. This effort aims to establish a favorable setting for the acceptance and assimilation of AI technologies in healthcare. Diwani and his colleagues are leading the way in utilizing AI to transform healthcare delivery and address intricate healthcare issues in Africa and beyond.

Godbless G. Minja is an Assistant Lecturer in the Department of Computer Science and Engineering (DoCSE) at the University of Dodoma (UDOM) in Tanzania. He completed his BSc in Computer Science at the University of Dar es Salaam (UDSM) in Tanzania and MSc in Cyber Security at the University of Birmingham in the United Kingdom (UK). He is currently pursuing a PhD in Information and Communication Science and Engineering (ICSE) at the Nelson Mandela African Institution of Science and Technology (NM-AIST) in Tanzania.

Kulwa Malyango is a Research Assistant and Software Developer at the AI4D Research Lab. He has a degree in Computer Science from the University of Dodoma in Tanzania. His research interests are in the application of artificial intelligence in the digital economy. Currently pursuing a master's degree in Computer Science, Kulwa is expanding his expertise in software development and artificial intelligence by working with esteemed researchers at the AI4D Research Lab. His main goal is to contribute meaningful research and practical solutions that can positively impact the digital economy, both locally and globally. He envisions a future where AI technologies are responsibly integrated into various sectors to improve people's lives.

Preface

Education is the most powerful weapon that you can use to change the world.
—Nelson Mandela

Machine learning is evolving rapidly, and its impact on our lives is profound. Machine learning applications have seamlessly expanded beyond their initial domains, integrating into our daily lives in ways we might not consciously recognize. It is not always apparent that machine learning algorithms drive commonplace activities such as using virtual assistants for voice commands, relying on self-driving features in modern vehicles, benefiting from smart home devices like intelligent kitchen appliances, or even experiencing personalized recommendations during online transactions. Machine learning is not just a tool, it is a force that shapes our future.

Acquiring machine learning knowledge and skills is crucial for staying relevant and unlocking diverse career opportunities in, for example, agriculture, healthcare, engineering, and finance industries. The acquired skills are in high demand, offering a lucrative career path with personal growth. Furthermore, machine learning contributes to efficiency, creativity, and competitive advantages in business creation and optimization.

This book is a humble attempt to demystify the complexities of machine learning while emphasizing the crucial role of ethics in this transformative field. Embarking on a journey into machine learning can be both thrilling and daunting. This book serves as a guide, simplifying concepts and providing practical examples to make the learning process engaging and accessible. This book is tailored to varied readers, including students, professionals exploring a new domain, or simply curious about the intersection of machine learning and ethics.

In the following chapters, this book will delve into the fundamentals of machine learning, explain the underlying algorithms, and explore real-world use cases of machine learning. The ethical implications of AI are central to the discussions in this book. As we unlock the potential of machine learning, we must also grapple with the responsibility it places on our shoulders. Notably, the ethical dimension of machine learning cannot be overstated. This book navigates the ethical considerations inherent in designing, deploying, and using machine learning models. From bias in algorithms to the societal impact of automation, urging the readers to think critically and responsibly about the power they wield as practitioners in this field. We encourage the reader to read the book with curiosity, an open mind, and a keen interest in the ethical dimensions of artificial intelligence. May this book empower the reader to delve into this transformative field responsibly and ethically.

Acknowledgments

This book was prepared by the AfriAI Research Lab with the aid of a grant from the AI for Development in Africa Program, a program funded by Canada's International Development Research Centre (IDRC), Ottawa, Canada and the Swedish International Development Cooperation Agency (Sida).

The authors wish to extend their heartfelt appreciation to the University of Dodoma (UDOM) and the Nelson Mandela African Institution of Science and Technology (NM-AIST) for the support they extended to the authors during the preparation of this book. Moreover, the authors acknowledge the support from other colleagues in the lab, as well as all individuals and institutions not explicitly mentioned here, who have contributed to the accomplishment of this book. Your invaluable contributions and efforts have been highly beneficial and are greatly appreciated.

The authors would also like to express their sincere gratitude to Dr. Tegawende Bissyande, Prof. Solomon Sunday Oyelere, and Dr. Ismaila Sanusi, who carefully reviewed the early drafts of this book and suggested various improvements.

Finally, the authors thank their families for the wholehearted support they devoted throughout the preparation of this book.

Glossary

Algorithm: A set of instructions that a computer follows to solve a problem.
Artificial Intelligence: Mimicking human intelligence in machines designed to perform tasks that usually require human intelligence.
Bias: A systematic error in a machine learning model that causes it to make incorrect predictions.
Data Preprocessing: Preparing and cleaning raw data before feeding it into a machine learning algorithm.
Feature Extraction: The process of selecting the most relevant features from the input data to enable efficient and accurate learning.
Hyperparameters: Parameters that are not learned from the data but are set manually before training a machine learning model.
Machine Learning: A branch of artificial intelligence that focuses on developing algorithms and models that enable computers to learn from data and make predictions without being explicitly programmed.
Metadata: A set of details that provides information about the data, such as data generation date, data source, data size, owner, and license agreement.
Model Development: Designing, building, and refining a machine learning model to solve a specific problem or make predictions.
Model Deployment: Deploying a trained machine learning model into a production environment.
Model Evaluation: The process of assessing the performance of a machine learning model using various metrics to determine its effectiveness.
Neural Network: A network of artificial neurons or nodes that draws inspiration from the structure and function of the human brain.
Noise: The presence of inaccurate or irrelevant variations in a dataset.
Overfitting: A phenomenon in machine learning where a model learns excessively well from the training data and fails to generalize to new, unseen data.
Outlier: A single data point in a dataset that deviates noticeably from the rest of the dataset.
Supervised Learning: A type of machine learning where a model is trained using labeled data, meaning each data point has a corresponding target value. The model learns to predict these target values for new, unseen data.
Testing Data: A set of data (different from the training data) used to evaluate the performance and generalization of a trained machine learning model.
Training Data: The labeled or unlabeled data used to train a machine learning model.

Underfitting: A phenomenon that occurs when a machine learning model fails to capture the underlying patterns in the training data, resulting in poor performance on both the training and testing data.

Unsupervised Learning: A type of machine learning where the model learns from unlabeled data, discovering hidden patterns and structures within the data without predefined target values.

Validation Set: A set of data (different from the training and testing datasets) that is used to fine-tune the performance of a machine learning model.

Fundamentals of machine learning

Upon completing this chapter, learners should be able to:
1. Define machine learning with a foundational understanding of its principles, terminologies, and processes.
2. Articulate the importance of machine learning, its practical applications, and growing relevance.
3. Differentiate between various types of machine learning algorithms, their characteristics, and use cases.
4. Examine real-world applications of machine learning across diverse industries and their practical impact.
5. Understand the interdisciplinary connections of machine learning with other computer science disciplines.

1.1 WHAT IS MACHINE LEARNING?

Machine learning is a field of science that utilizes data and algorithms to train computers to mimic human learning processes, as illustrated in Figure 1.1. It involves learning from data to acquire knowledge (i.e., what is learnt), understand the process (i.e., how it learnt), and apply this knowledge to solve problems (i.e., reasoning and decision-making) reliably.

Additionally, as summarized in Figure 1.2, machine learning can also be defined as the science of creating autonomous software or models that learn from data to solve problems and make predictions. Simply put, machine learning focuses on building models that improve automatically with experience. This approach offers greater flexibility and efficiency, significantly reducing software developers' need to manually program machine instructions.

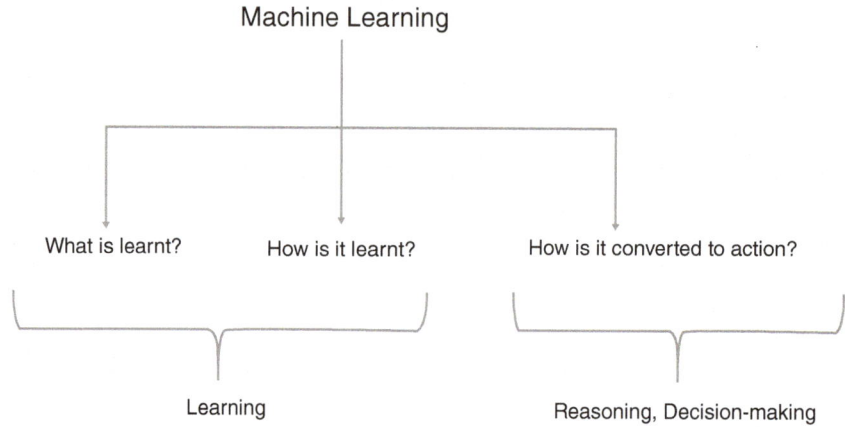

FIGURE 1.1 Machine learning overview.

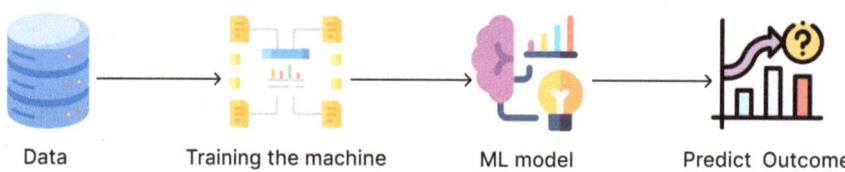

FIGURE 1.2 The meaning of machine learning.

1.2 A BRIEF HISTORY OF MACHINE LEARNING

Machine learning has been evolving since its inception in the 1950s. In the 1970s and 1980s, the field of machine learning primarily revolved around the goal of decision-making based on predetermined rules. However, in the 1990s, a notable shift occurred, redirecting machine learning toward a more data-centric approach. During the 2000s, there was a significant advancement in computer learning capabilities, particularly complex and data-rich applications, for example, processing visual information. This progress greatly contributed to the machines' ability to learn and comprehend, mirroring the way human brains work. In the 2010s, machine learning experienced remarkable progress, marked by significant developments in voice assistants, self-driving technologies, recommendation systems, and the widespread adoption of spam filters and chatbots. In the early 2020s, ongoing trends include the exploration of federated learning, allowing model training across decentralized machines and an increased focus on ethical considerations. Additionally, machines can learn and generate content in human-like language and create original and creative outputs on their own, such as images, text, music, or even entire realistic scenarios. Furthermore, the field continues to evolve, emphasizing responsible AI practices, bias mitigation, and the development of models that align with ethical principles.

1.3 TYPES OF MACHINE LEARNING ALGORITHMS

An algorithm is a set of mathematical instructions or rules that directs a computer program to solve a specific problem or perform a task. In machine learning, an algorithm enables a model to process data, identify patterns, and make predictions. It is the fundamental building block that drives the learning process and allows the model to generalize its knowledge to new, unseen data. There are four main types of machine learning algorithms: Supervised Learning, Unsupervised Learning, Semi-supervised Learning, and Reinforcement Learning, which are discussed in the following subsections.

1.3.1 Supervised learning

Supervised learning is a branch of machine learning wherein the algorithm learns from input features associated with known output labels or target values, enabling it to predict or classify new, unseen data. Supervised learning relies on a dataset containing input-output pairs to train the algorithm. This concept can be likened to learning under the guidance of a supervisor. Generally, supervised learning proves more effective when labeled datasets are available than other learning methods. Its applications span various real-life scenarios, including fraud detection (e.g., distinguishing between fraudulent and legitimate transactions), sales forecasting (e.g., predicting high, medium, or low sales), and email categorization (e.g., identifying spam emails). Table 1.1 depicts an example of a labeled dataset containing the diabetes diagnostic measurements, where the last feature (label) contains the values of 1 or 0, indicating that a patient is diabetic or not, respectively.

1.3.1.1 Types of supervised learning

Supervised learning encompasses two primary types of problems: classification and regression.

- **Classification** entails assigning input data samples into predefined categories or classes. Drawing from previous learning experiences, a classification algorithm typically identifies data samples within the input dataset

TABLE 1.1 An example of a labeled dataset

INDEX	PREGNANT	GLUCOSE	BP	SKIN	INSULIN	BMI	PEDIGREE	AGE	LABEL
0	6	148	72	35	0	33.6	0.627	50	1
1	1	85	66	29	0	26.6	0.351	31	0
2	8	183	64	0	0	23.3	0.672	32	1

and assigns them to specific classes. Common classification types include binary and multiclass classification.
- In binary classification, an algorithm trains to classify data samples into one of two potential classes, aiming to construct a model capable of accurately assigning new samples to their respective classes. Examples of binary classification applications include discerning whether an email is spam or not and diagnosing whether an individual is diabetic or not. Algorithms proficient in binary classification include Logistic Regression, Support Vector Machine, k-Nearest Neighbors, Decision Trees, Naive Bayes, and Random Forest.
- In multiclass classification, an algorithm trains to classify data samples into three or more classes, aiming to construct a model capable of accurately categorizing new data samples into their respective classes. Examples of multiclass classification applications include determining the genre of a movie into categories such as action, drama, comedy, or fiction and classifying animals into categories like dog, cat, or tiger. Algorithms proficient in multiclass classification include Decision Tree and Artificial Neural Networks.
- **Regression** involves predicting a continuous output or numerical value based on input features. To develop a model, regression algorithms are trained to understand the relationship between independent variables and a continuous dependent variable. This model can then predict the outcome of new, unseen input data. An example of a regression algorithm is Linear Regression.
  - In linear regression, an algorithm captures the relationship between a dependent variable (the feature to be predicted) and one or more independent variables (the predictor features) to develop a model capable of accurately predicting the dependent variable based on at least one independent variable. Linear regression algorithms are categorized into two main types: simple and multivariate (multiple) regression. Simple linear regression involves a dependent variable relying on a single independent variable, while multivariate linear regression involves a dependent variable relying on multiple independent variables. Linear regression is extensively used for tasks such as estimating housing prices based on factors like area, room count, and location.

1.3.2 Unsupervised learning

Unsupervised learning is a type of machine learning that does not require labeled datasets. Instead of being guided by predefined labels, the algorithm independently discovers hidden patterns and insights within the data. Unsupervised learning is crucial because obtaining unlabeled data is often easier than acquiring labeled data, which typically requires human annotators. Additionally, unsupervised learning can help identify features useful for categorization. Table 1.2 illustrates an unlabeled dataset containing diabetes diagnostic measurements without a label feature.

TABLE 1.2 An example of an unlabeled dataset

INDEX	PREGNANT	GLUCOSE	BP	SKIN	INSULIN	BMI	PEDIGREE	AGE
0	6	148	72	35	0	33.6	0.627	50
1	1	85	66	29	0	26.6	0.351	31
2	8	183	64	0	0	23.3	0.672	32

1.3.2.1 Types of unsupervised learning

Unsupervised learning techniques are categorized into clustering and association rules as described in the following:

1.3.2.1.1 Clustering
This machine learning technique finds patterns or structures in a collection of unclassified data and uses them to group similar data into clusters or segments. Common categories of clustering algorithms include hierarchical, partitioning, and density-based clustering.

- Hierarchical clustering involves creating a hierarchical structure of clusters by merging or splitting clusters based on the similarity of data points. The application of hierarchical clustering spans various domains, including document clustering and social network analysis. Notable examples of hierarchical clustering algorithms include agglomerative hierarchical clustering, divisive hierarchical clustering, and Ward's method.
- Partitioning clustering algorithms organize a dataset into distinct, non-overlapping groups or clusters, where each data point belongs to only one cluster. Partitioning clustering is used in applications such as customer segmentation based on online purchasing behavior. Examples of partitioning clustering algorithms include K-means, fuzzy C-means (FCM), X-means, and G-means.
- Density-based clustering algorithms cluster data points according to their density in the feature space. The algorithm identifies clusters as regions with a higher density of data points, separated by areas of lower density. This enables it to uncover clusters of diverse shapes and effectively handle noise or outliers. In density-based clustering, clusters emerge around dense regions, while data points in sparser regions may be classified as outliers. Density-based clustering algorithms find applications in traffic analysis and anomaly detection in network security. Examples of such algorithms include Density-Based Spatial Clustering of Applications with Noise (DBSCAN) and mean shift.

1.3.2.1.2 Association Rules
This technique is used to identify relationships or associations between variables in a dataset based on predefined rules. The rules highlight patterns in the form of "*if-then*" statements, indicating that the occurrence of one set of items is associated with

the occurrence of another set of items. Applications of association rules are market basket analysis, online shopping customer behavior analysis, and inventory management. In essence, association rules provide valuable insights into the relationships between seemingly unrelated data points, facilitating data-driven decision-making in diverse fields. For instance, businesses utilize association rules to understand patterns of co-occurrence among products frequently purchased together (e.g., bread and jam, book and pencil), informing decisions on product placement, targeted marketing, and personalized recommendations. Examples of association rules algorithms are Apriori, Eclat, and FP-Growth.

1.3.3 Semi-supervised learning

Semi-supervised learning provides the capability to train an algorithm using a combination of labeled data, consisting of a small number of examples with known labels, and unlabeled data, which comprises a large number of examples without labels. In situations where acquiring fully labeled data is challenging, unsupervised and semi-supervised learning offers viable alternatives to supervised learning. The process of creating labeled datasets can be time-consuming, labor-intensive, and costly, as it often requires the involvement of domain experts for manual annotation. Various algorithms can be employed in semi-supervised learning, including self-training, co-training, generative models, entropy regularization, graph-based methods, semi-supervised support vector machines (S3VM), and transductive support vector machines (SVM).

1.3.4 Reinforcement learning

Reinforcement learning involves training a machine learning model to make a series of decisions within a complex environment. The model perceives and interprets its surroundings, employing a trial-and-error approach to discover the optimal solution to a given problem. In reinforcement learning, the model receives rewards for desirable behaviors and may face penalties for undesired ones. Several common reinforcement learning algorithms exist, including Q-learning, deep Q-networks (DQN), policy gradient methods, actor-centric methods, Monte Carlo methods, and deep deterministic policy gradient (DDPG).

1.4 RELATIONSHIP BETWEEN MACHINE LEARNING AND OTHER COMPUTER SCIENCE DISCIPLINES

This section describes the relationship between machine learning and artificial intelligence, data science, traditional programming, deep learning, natural language processing, computer vision, and generative AI.

1.4.1 Machine learning and artificial intelligence

In brief, AI is a field within computer science, with machine learning under its umbrella. While machine learning and AI are often used interchangeably, machine learning is a subset of AI that enables systems to learn and refine processes without explicit programming for each task. These systems ingest data, process it through algorithms, and learn from it, discerning patterns or anomalies. In contrast, AI involves crafting systems to think and behave in ways akin to humans, empowering them to undertake tasks typically requiring human intellect. AI is characterized in two key ways: as the scientific endeavor to design machines capable of decision-making like humans and as the manifestation of intelligence in machines, distinct from natural human and animal intelligence. In essence, machine learning outputs contribute to AI solutions. While both share similar goals and functions, AI covers various techniques such as computer vision, natural language processing, and robotics.

1.4.2 Machine learning and data science

Data science is a discipline that revolves around studying data and extracting valuable insights from it. On the other hand, machine learning is a specialized field within data science that focuses on comprehending and constructing models that leverage data to enhance performance or make predictions. In simpler terms, data science aims to extract actionable insights from data, while machine learning is concerned with developing models that can automate predictive behavior by utilizing the available data. The relationship between machine learning, AI, and data science is illustrated in Figure 1.3.

1.4.3 Machine learning and traditional programming

Both machine learning and traditional programming serve as problem-solving tools, each suitable for different types of challenges. Traditional programming excels in scenarios with well-defined rules and structures, where solutions can be articulated

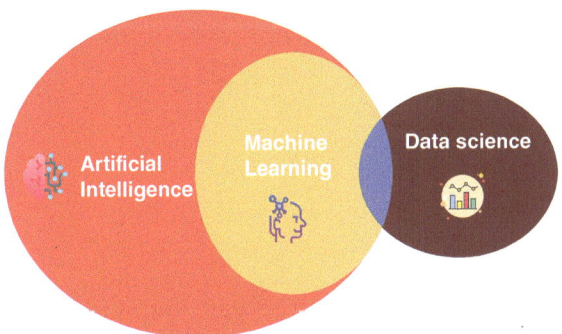

FIGURE 1.3 The relationship between AI, machine learning, and data science.

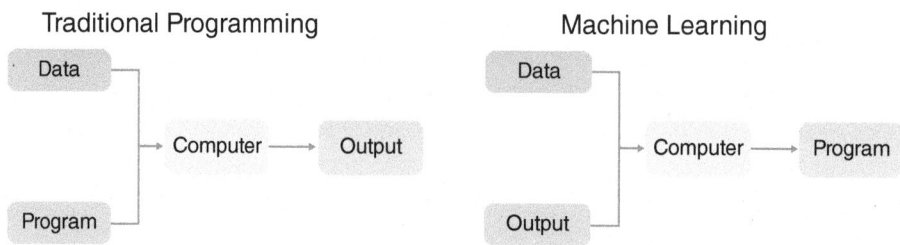

FIGURE 1.4 The relationship between machine learning and traditional programming.

through logical statements and algorithms. Conversely, machine learning shines in addressing problems characterized by complex and elusive patterns or relationships. Inspired by human learning, machine learning empowers computers to glean insights from examples and autonomously devise solutions. As depicted in Figure 1.4, traditional programming involves the computer processing data and programs to generate an output. In contrast, machine learning entails the computer utilizing data and expected output to generate the program.

1.4.4 Machine learning and deep learning

Both machine learning and deep learning reside under the umbrella of artificial intelligence, yet they diverge in their learning methods and problem-solving approaches. Deep learning, a subset of machine learning, is distinguished by its utilization of neural networks, inspired by the human brain, to learn and solve problems. In contrast, machine learning trains computer programs or systems to execute tasks without explicit instructions. Machine learning excels in well-defined tasks with structured and labeled data, typically involving lower data volumes. Conversely, deep learning thrives in tackling complex tasks with unstructured and extensive data. Examples of machine learning applications encompass spam filtering, image recognition, and product recommendation systems. In contrast, deep learning finds applications in self-driving cars, speech recognition, medical image analysis, and generative AI applications like chatbots (such as ChatGPT) and Google's Gemini. Figure 1.5 delineates the relationship between artificial intelligence, machine learning, and deep learning.

1.4.5 Machine learning and natural language processing

Natural language processing (NLP) is a specialized field within machine learning that focuses on the interaction between human language and computers. It recognizes the abundance of valuable information in text and speech data, such as news articles, customer reviews, and research papers. NLP provides computational tools to extract insights and derive meaning from this unstructured data, making it a crucial component of the machine learning toolbox for understanding and processing human language.

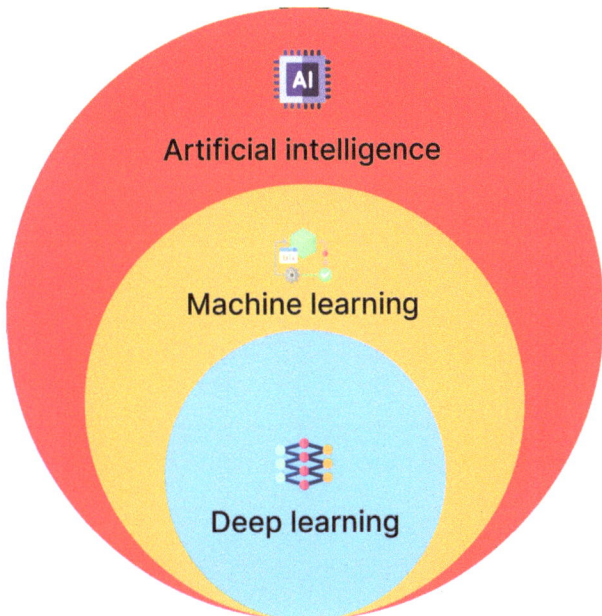

FIGURE 1.5 The relationship between artificial intelligence, machine learning, and deep learning.

The applications of NLP are diverse and span across various industries. In healthcare, NLP can be used for tasks like clinical text analysis and medical record extraction. In education, it can aid in automated grading and intelligent tutoring systems. Communication platforms like Google Translate and text auto-completion rely on NLP algorithms. In business and marketing, sentiment analysis and chatbots employ NLP techniques. Also, NLP contributes to entertainment applications such as social media feed recommendations and voice assistants. Figure 1.6 illustrates the relationship between machine learning and natural language processing, highlighting how NLP plays a vital role in leveraging machine learning techniques to process and understand human language.

1.4.6 Machine learning and computer vision

Computer vision (CV) constitutes a subset of AI that empowers computers to comprehend visual information such as images and videos. Given the complex and variable nature of visual data, traditional programming techniques often fall short in resolving many computer vision tasks. Instead, machine learning methods, particularly deep learning, are leveraged to discern visual patterns from images autonomously. This progression underlies the creation of applications like image classification (categorizing images), object detection (locating specific objects within images), and facial recognition (matching and identifying human faces). The relationship among artificial intelligence, machine learning, and computer vision is succinctly depicted in Figure 1.7.

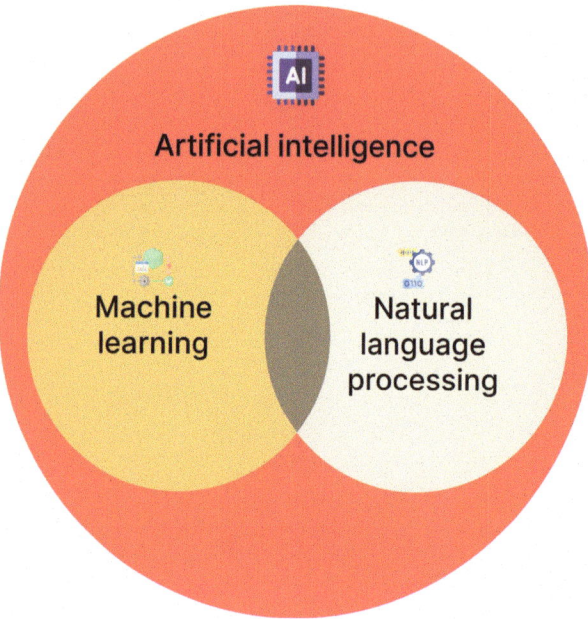

FIGURE 1.6 The relationship between machine learning and natural language processing.

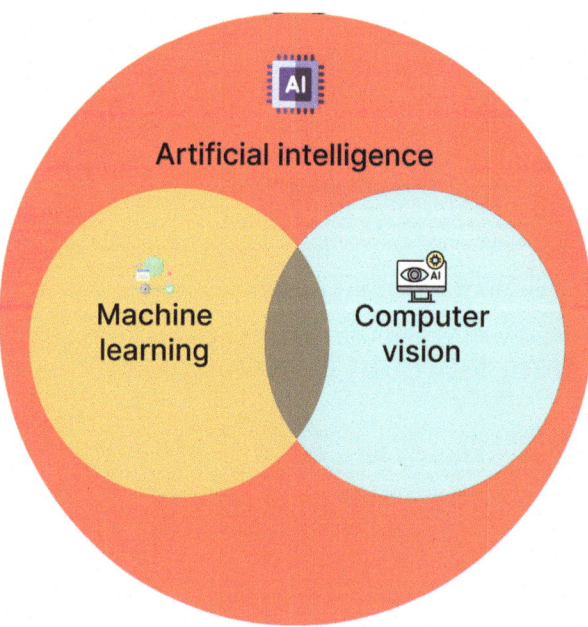

FIGURE 1.7 The relationship between artificial intelligence, machine learning, and computer vision.

1.4.7 Machine learning and generative AI

AI has made significant strides in recent years, showcasing mastery in various domains, from complex games to language translation and disease diagnosis. However, what if AI could transcend its current capabilities and become a creator? You may have encountered ChatGPT, a chatbot with human-like conversational abilities, or Midjourney, a model capable of generating realistic images from textual prompts. These breakthroughs are powered by Generative AI, a subset of machine learning, particularly deep learning, which focuses on generating novel content rather than just analyzing or acting upon existing data. The following key advancements have propelled the evolution of Generative AI:

i. **Transformers**: These architectures revolutionized NLP, enabling AI systems to understand the relationships between words and the language context more sophisticatedly. This paved the way for advanced conversational AI applications.
ii. **GANs (Generative Adversarial Networks)**: These systems operate within a framework where two deep neural networks engage in a competitive process: a generator network strives to produce realistic data, while a discriminator network differentiates between real and generated examples. This dynamic competition fosters a cycle of continuous enhancement, driving improvements in the quality and realism of the generated content.
iii. **Diffusion Models**: These learn to create new data by gradually reversing a process of adding noise to existing data. They have proven exceptionally powerful in generating high-resolution images and other complex media.

1.5 THE IMPORTANCE OF MACHINE LEARNING

Machine learning models streamline tasks that would typically demand manual effort from humans. By harnessing machine learning, organizations can uncover valuable insights from data, facilitating informed decision-making processes. Implementing data-driven strategies enhances business efficiency, performance, and productivity and mitigates risks. The significance of machine learning extends across various sectors and industries, enabling the anticipation of future risks and opportunities. In healthcare, for instance, machine learning can scrutinize medical images, genomic data, and electronic health records to aid physicians in making precise diagnoses and recommending suitable treatments. Similarly, machine learning finds utility in crop monitoring, yield prediction, pest detection, and soil analysis in agriculture. Machine learning optimizes crop production and resource allocation by enabling farmers to make data-driven decisions.

1.6 WHEN DO WE NEED MACHINE LEARNING?

Machine learning is indispensable across various contexts and problem domains, especially where conventional rule-based programming or manual analysis falls short. The following are several typical scenarios where machine learning proves exceptionally beneficial:

a. **Handling of large and complex data**: Machine learning algorithms excel at revealing patterns, correlations, and insights that are challenging to discern manually, especially when confronted with extensive datasets. With its capacity to navigate complex data structures and high-dimensional data, machine learning is well-suited for data mining, pattern recognition, and predictive modeling tasks.

b. **Need for automation and efficiency**: Machine learning can automate repetitive tasks and boost efficiency across various domains. In customer service, for instance, integrating chatbots driven by machine learning can efficiently handle basic inquiries, allowing human agents to focus on more complex issues.

c. **Prediction and forecasting**: Machine learning algorithms can analyze historical data patterns to forecast outcomes across various sectors, including stock price movements, weather patterns, and disease outbreaks. By harnessing this capability, machine learning provides invaluable insights, enabling informed decision-making grounded in past trends.

d. **Anomaly detection**: Machine learning algorithms can detect anomalies and recognize unusual patterns within data. This capability holds significant value across various domains, including fraud detection, cybersecurity, and network monitoring. By acquiring knowledge of normal behavior through training, machine learning models can effectively identify deviations and anomalies that may indicate fraudulent activities or security breaches. This enables prompt intervention and implementation of mitigation measures to address potential risks and safeguard the system or network.

e. **Personalization and recommendation systems**: Machine learning empowers personalized experiences and tailored recommendations by scrutinizing user preferences and behavior. This technology drives recommendation engines across diverse domains like hospitality, content streaming, and social media. By analyzing user data, machine learning models deliver personalized suggestions for products, services, movies, or connections that resonate with individual preferences.

f. **Computational linguistics**: Machine learning is pivotal in computational linguistic tasks, empowering machines to process, comprehend, and interpret human languages effectively; this involves language translation, sentiment analysis, speech recognition, and chatbots. Through extensive training on vast amounts of text data, machine learning models can grasp and generate human language, facilitating seamless communication and language-based interactions.

1.7 MACHINE LEARNING SKILLS

Given the interdisciplinary nature of machine learning, the requisite skills lie at the intersection of various domains, including software engineering, data science, and communication. These skills can be broadly categorized into technical and soft skills, as elaborated in the following subsections.

1.7.1 Essential technical skills for machine learning professionals

This refers to the technical skills spanning data science and software engineering, as summarized in Table 1.3.

1.7.2 Essential soft skills for machine learning professionals

Soft skills are what set apart effective machine learning professionals from those who are ineffective. These skills are needed for the project's successful completion and delivery. Such skills include communication, problem-solving, time management, teamwork, and a thirst for learning.

TABLE 1.3 Essential technical skills for machine learning professionals

SKILL	DESCRIPTION
Software Engineering	Includes the ability to write computer programs, understanding of algorithms and data structures, and knowledge of computer architecture and organization.
Statistics and Mathematics	This entails having proficiency in hypothesis testing, data modeling, and a strong grasp of mathematical concepts such as probability, statistics, and linear algebra. It also involves the ability to devise an evaluation strategy for predictive models and algorithms.

1.8 WHAT DO MACHINE LEARNING PROFESSIONALS DO?

Machine learning professionals are responsible for designing, building, testing, deploying, and updating machine learning models. In particular, this involves:

- Performing data analysis.
- Running machine learning experiments.
- Implementing machine learning models.
- Optimizing the machine learning models.
- Deploying machine learning models into production.

Additionally, a significant aspect of the role involves collaborating with various stakeholders, including domain experts, data scientists, researchers, software engineers, and product managers, to establish project objectives and roadmaps.

1.9 REAL-WORLD APPLICATIONS OF MACHINE LEARNING

Machine learning finds applications across numerous domains: manufacturing, retail, healthcare and life sciences, transportation, digital economy, agriculture, environmental conservation, and education. Table 1.4 provides real-world examples of machine-learning applications in various fields.

1.10 MACHINE LEARNING AND ETHICAL CONCERNS

The ethical considerations surrounding machine learning are increasingly paramount as the technology progresses. Issues such as bias, explainability, privacy, transparency, algorithmic fairness, safety, job displacement, and weaponization necessitate a comprehensive approach. Prioritizing fairness, accountability, and transparency entails investing in pertinent research, crafting supportive ethical frameworks, and implementing requisite policies and regulations. These endeavors are vital to ensure that the potential benefits of machine learning outweigh its potential harms. Moreover, raising public awareness regarding the ethical implications of machine learning is pivotal in fostering conscientious and informed utilization of this technology. Addressing these concerns collectively will guarantee that machine learning evolves and is deployed to align with societal values and ethical standards, ultimately serving the greater good.

TABLE 1.4 A summary of some real-world applications of machine learning

APPLICATION	DESCRIPTION
Image recognition	Identification and classification of objects or patterns within digital images. Application examples include labeling an X-ray image as cancerous or not and assigning a name to a photographed face (this is known as "tagging" on Facebook).
Speech recognition	Translating speech into a readable text that the machine can understand and work on. This results in applications capable of responding to speech. Speech recognition is used for voice search and dialing, and application control. Real-world speech recognition applications include Google Home, Google Assistant, Alexa, Siri, and Cortana.
Medical diagnosis	Studying physiological data, environmental influences, and genetic factors complements the decision-making by medical doctors to diagnose diseases early and effectively. Examples of real-life applications include Dr. Elsa[a], CareAi, and Ada Health[b].
Agriculture	Enabling accurate and efficient farming with less manpower for high and quality yields. It can be used to predict crop yield as well as detect and assess the impact of crop diseases. Application examples include Plantix[c], Trace Genomics, and Agrio[d].
Automotive industry	Building self-driving cars integrated with various models and algorithms that analyze data collected from cameras and sensors, interpreting them, and making decisions accordingly. Common examples include Google's and Tesla's self-driving cars.
Travel assistance	Virtual travel agents that enhance the overall travel experience for users. Examples of real-life applications include Google Maps, commercial flights, and riding apps like Uber and Bolt.
Entertainment	Recommending personalized entertainment content based on the user's history. For example, Netflix recommends movies based on users' past behaviors. Facebook gathers behavioral information for every user on social media platforms and uses it to predict interests and recommends articles and notifications on news feed.
Email Intelligence	Enhancing intelligence capabilities of email applications. Examples include email classification (e.g., spam filtering) and smart replies.
Cyber security	Detecting and preventing security threats. Machine learning applications in cybersecurity include intrusion detection, malware detection, anomaly detection, vulnerability detection, and fraud detection.
Surveillance	Analyzing video or image data for object detection, tracking, and behavior recognition. Examples of applications are video surveillance, crowd monitoring, and real-time alert and response.

Notes:
[a] https://www.elsa.health/
[b] https://ada.com/
[c] https://plantix.net/en/
[d] https://agrio.app/

1.11 SUMMARY

This chapter introduces the fundamental concepts of machine learning, its relationship with other related concepts, and its overall significance. Furthermore, the chapter explores various scenarios in which machine learning is essential. It also presents the crucial skills required for professionals in machine learning. Real-world examples are offered, showcasing the practical applications of machine learning and highlighting its relevance and impact. Different types of machine learning problems are discussed, and lastly, the chapter concludes by briefly highlighting the ethical concerns of machine learning.

Exercises

1. With examples of any three industries in which machine learning is used, give thorough descriptions of how it is used.
2. Give descriptions of what machine learning models promise in software development.
3. With at least three algorithm examples for each, provide thorough descriptions of the four main types of machine learning algorithms (and their respective sub-types where applicable).
4. Provide descriptions of the relationship between machine learning and the following disciplines:
   a. Artificial Intelligence
   b. Data Science
   c. Traditional Programming
   d. Deep Learning
5. In detail, describe the following terminologies:
   a. Generative AI
   b. NLP
6. Explain the importance of machine learning.
7. With examples, outline the scenarios in which machine learning is needed.
8. Provide descriptions of the essential skills for machine learning professionals.
9. Explain any five real-world machine learning applications with at least two examples for each.
10. Briefly explain the ethical concerns of machine learning.

FURTHER READING

Dönmez, P. (2013). Introduction to machine learning, by Ethem Alpaydın. Cambridge, MA: The MIT Press2010. ISBN: 978-0-262-01243-0. *Natural Language Engineering*, 19(2), 285–288.

Firican, G. (2023). *The history of machine learning*. Retrieved December 12, 2023, from https://www.lightsondata.com/the-history-of-machine-learning/

Gonsalves, T. and Upadhyay, J. (2021). Integrated deep learning for self-driving robotic cars. *AI for Future Generation Robotics*, Elsevier, pp. 93–118.

Gupta, S. (2021, June 14). What skills do you need to become a Machine Learning engineer? *Springboard*. https://www.springboard.com/blog/data-science/machine-learning-skills/

Lateef, Z. (2021, December 19). Introduction to machine learning: All you need to know about machine learning. *Edureka*. https://www.edureka.co/blog/introduction-to-machine-learning/

Li, Y. F., & Liang, D. M. (2019). Safe semi-supervised learning: A brief introduction. *Frontiers of Computer Science*, 13(4), 669–676.

Oracle. (2022). What is big data? Retrieved July 7, 2022, from https://www.oracle.com/in/big-data/what-is-big-data/

Pan, S. J., & Yang, Q. (2009). A survey on transfer learning. *IEEE Transactions on Knowledge and Data Engineering*, 22(10), 1345–1359.

Rice University. (2022). Computer science Vs. Artificial intelligence/machine learning: What's the difference? Retrieved July 7, 2022, from https://csweb.rice.edu/academics/graduate-programs/online-mcs/blog/computer-science-vs-artificial-intelligence-and-machine-learning

Salesforce Blog. (2022). Machine learning: 6 real-world examples. Retrieved July 8, 2022, from https://www.salesforce.com/eu/blog/2020/06/real-world-examples-of-machine-learning.html

Szepesvári, C. (2010). Algorithms for reinforcement learning. *Synthesis Lectures on Artificial Intelligence and Machine Learning*, 4(1), 1–103.

Yale University. (2022). Machine learning. Retrieved July 7, 2022, from https://cpsc.yale.edu/research/machine-learning

Mathematics for machine learning

2

Upon completing this chapter, learners should be able to:

1. Understand basic mathematics essential for comprehending machine learning concepts.
2. Master the concept of representing machine learning models mathematically, enabling good understanding and implementation.
3. Develop the capability of converting machine learning problems into formulations for mathematical optimization.
4. Understand the working principles of machine learning algorithms by analyzing and comprehending mathematical expressions.
5. Apply mathematical representations to assess algorithmic performance, model behavior, and problem-solving capability in machine learning contexts.

2.1 LINEAR ALGEBRA

Linear algebra is a fundamental component of mathematics that is essential for machine learning practitioners. It provides the theoretical foundation needed to understand and work with various machine learning concepts. Mastery of linear algebra equips learners with the critical tools and arithmetic computations required for implementing and optimizing machine learning algorithms. The following subsections present in detail scalars, vectors, matrices, eigenvalues, and eigenvectors which are considered to be the basic concepts of linear algebra.

2.1.1 Scalars

In mathematics, a scalar is a measurement that has a magnitude without any associated direction. Within the era of machine learning or data science, scalars might represent various features of data points. For instance, residence datasets with the following features: number of bedrooms, the total floor area, and the sale price of each house can be represented as separate scalar numbers. Scalar values are fundamental units used to create more complex mathematical models and are crucial for carrying out mathematical computations and analyses in machine learning algorithms. Scalars cover various numerical values such as integers, decimals, fractions, and irrational numbers. However, depending on their importance, scalars can be either positive, negative, or zero. Scalars can be evaluated in mathematics using standard arithmetic operations such as addition, subtraction, multiplication, and division.

For example, consider two scalars, $a = 5$ and $b = 3$. The sum of these two scalars is obtained by adding them together, $a + b = 5 + 3 = 8$.

2.1.2 Vectors

A vector is a collection of numbers that are ordered consecutively. However, vectors are quantities that can convey direction as well as magnitude. Equation (2.1) depicts this concept, which can be identified as a row or column of numbers in lowercase characters, such as v.

$$v = (v_1, v_2, v_3) \tag{2.1}$$

where v_1, v_2, v_3 are scalar values, often real values.

In mathematical operations, vectors can be calculated using standard arithmetic operations such as addition, subtraction, and multiplication, as discussed in the subsequent sections.

2.1.2.1 Vector addition

Consider two vectors; $a = (a_1, a_2, a_3)$ and $b = (b_1, b_2, b_3)$. Vector addition of "a" and "b" is performed element-wise to produce a new vector of the same length as shown in Equation (2.2).

$$a + b = (a_1 + b_1, a_2 + b_2, a_3 + b_3) \tag{2.2}$$

For example, let us say we have two vectors, $a = (2, 4, 6)$ and $b = (1, 3, 5)$. To find the sum of these vectors, corresponding components of the vectors will have to be added to each other, as shown in the following:

$$a + b - (2 + 1, 4 + 3, 6 + 5).$$

Thus $a + b$ is equal to (3, 7, 11)

2.1.2.2 Vector subtraction

Consider two vectors; $a = (a_1, a_2, a_3)$ and $b = (b_1, b_2, b_3)$. Vector subtraction of "a" and "b" is performed element-wise to produce a new vector of the same length as shown in Equation (2.3).

$$a - b = (a_1 - b_1, a_2 - b_2, a_3 - b_3) \qquad (2.3)$$

For example, let us say we have two vectors, $a = (2, 4, 6)$ and $b = (1, 3, 5)$. To subtract vector "b" from "a", the corresponding components will have to be subtracted from each other as shown in the following:

$$a - b = (2 - 1, 4 - 3, 6 - 5).$$

Thus $a - b$ is equal to (1, 1, 1)

2.1.2.3 Vector multiplication

It is worth noting that multiplication is typically defined for vectors of the same dimension when dealing with vectors. This is because certain operations, like the dot product and cross product, require vectors of the same dimensionality to be performed. Given two vectors $a = (a_1, a_2, a_3)$ and $b = (b_1, b_2, b_3)$ of equal length, the dot product and cross product of "a" and "b" are given in Equations (2.4) and (2.5), respectively.

$$a \cdot b = (a_1 \times b_1, \; a_2 \times b_2, \; a_3 \times b_3) \qquad (2.4)$$

$$a \times b = (a_2 \times b_3 - a_3 \times b_2, \; a_3 \times b_1 - a_1 \times b_3, \; a_1 \times b_2 - a_2 \times b_1) \qquad (2.5)$$

Using the same vectors as before, $a = (2, 4, 6)$ and $b = (1, 3, 5)$. The dot product and cross product of these two vectors are:

$$a \cdot b = (2 \times 1, \; 4 \times 3, \; 6 \times 5)$$

$a \cdot b = 2 + 12 + 30 = 44$. Thus, $a \cdot b$ is equal to 44.

$$a \times b = (4 \times 5 - 6 \times 3, \; 6 \times 1 - 2 \times 5, \; 2 \times 3 - 4 \times 1)$$

$$a \times b = (20 - 18, \; 6 - 10, \; 6 - 4) = (2, -4, 2).$$

Thus, $a \times b$ is equal to $(2, -4, 2)$.

2.1.3 Matrix

A matrix is a grid of numbers arranged in rows and columns. Each number in a matrix is called an element. In machine learning, matrices are used to organize data, with each row representing an individual item or sample and each column representing a feature of that item. In addition, matrices serve as the foundational representation for datasets in machine learning, facilitating efficient analysis and processing throughout the machine learning workflow. A matrix is usually denoted by an uppercase letter (e.g., A), and each element is referred to by its two-dimensional subscript of row (i) and column (j) such as a_{ij} as represented in Equation (2.6).

$$A = \begin{pmatrix} a_{11} & a_{12} & a_{13} \\ a_{21} & a_{22} & a_{23} \end{pmatrix} \qquad (2.6)$$

Similar to vectors, matrices can be manipulated using standard arithmetic operations such as addition, subtraction, and multiplication, as discussed in the subsequent sections. However, the division of a matrix can only be performed on each of its elements by a scalar value.

2.1.3.1 Matrix addition

Matrix addition involves adding together corresponding elements of two matrices with the same dimension to form a new matrix whose elements are the sum of the respective elements from the two matrices being added together. In other words, the items in the i-th row and j-th column of matrices A and B are added together to form a new matrix. Given matrices A and B in Equations (2.7) and (2.8), respectively, the result of adding the two matrices is shown in Equation (2.9).

$$A = \begin{pmatrix} a_{11} & a_{12} & a_{13} \\ a_{21} & a_{22} & a_{23} \end{pmatrix} \qquad (2.7)$$

$$B = \begin{pmatrix} b_{11} & b_{12} & b_{13} \\ b_{21} & b_{22} & b_{23} \end{pmatrix} \qquad (2.8)$$

$$A + B = \begin{pmatrix} a_{11} + b_{11} & a_{12} + b_{12} & a_{13} + b_{13} \\ a_{21} + b_{21} & a_{22} + b_{22} & a_{23} + b_{23} \end{pmatrix} \qquad (2.9)$$

For example, given matrices A and B their sum is calculated as follows:

$$A = \begin{pmatrix} 1 & 2 & 3 \\ 4 & 5 & 6 \end{pmatrix}$$

$$B = \begin{pmatrix} 2 & 4 & 6 \\ 8 & 10 & 12 \end{pmatrix}$$

$$A + B = \begin{pmatrix} 1+2 & 2+4 & 3+6 \\ 4+8 & 5+10 & 6+12 \end{pmatrix} = \begin{pmatrix} 3 & 6 & 9 \\ 12 & 15 & 18 \end{pmatrix}$$

2.1.3.2 Matrix subtraction

Matrix subtraction can be performed between two matrices with the same dimension and involves subtracting each element of the second matrix from its corresponding element of the first matrix to produce a new matrix. In other words, an element in the *i-th* row and *j-th* column of matrix B is subtracted from the corresponding element in the *i-th* row and *j-th* column of matrix A. Given matrices A and B in Equations (2.10) and (2.11), respectively, the result of subtracting matrix B from A is given as shown in Equation (2.12).

$$A = \begin{pmatrix} a_{11} & a_{12} & a_{13} \\ a_{21} & a_{22} & a_{23} \end{pmatrix} \quad (2.10)$$

$$B = \begin{pmatrix} b_{11} & b_{12} & b_{13} \\ b_{21} & b_{22} & b_{23} \end{pmatrix} \quad (2.11)$$

$$A - B = \begin{pmatrix} a_{11} - b_{11} & a_{12} - b_{12} & a_{13} - b_{13} \\ a_{21} - b_{21} & a_{22} - b_{22} & a_{23} - b_{23} \end{pmatrix} \quad (2.12)$$

For example, given matrices A and B the subtraction of matrix B from A is calculated as follows:

$$A = \begin{pmatrix} 3 & 5 & 7 \\ 2 & 4 & 6 \end{pmatrix}$$

$$B = \begin{pmatrix} 1 & 2 & 3 \\ 1 & 2 & 3 \end{pmatrix}$$

$$A - B = \begin{pmatrix} 3-1 & 5-2 & 7-3 \\ 2-1 & 4-2 & 6-3 \end{pmatrix} = \begin{pmatrix} 2 & 3 & 4 \\ 1 & 2 & 3 \end{pmatrix}$$

2.1.3.3 Matrix multiplication

Matrix multiplication involves performing the dot product of the rows and columns of the multiplied matrices. In multiplying two matrices, each element in the resulting matrix is calculated by taking the dot product of the corresponding row of the first matrix and the corresponding column of the second matrix. This process repeats for each element in the resulting matrix. Given matrices A and B in Equations (2.13) and (2.14), respectively, the result of multiplying matrices A and B is given in matrix C as shown in Equation (2.15).

$$A = \begin{pmatrix} a_{11} & a_{12} \\ a_{21} & a_{22} \end{pmatrix} \quad (2.13)$$

$$B = \begin{pmatrix} b_{11} & b_{12} \\ b_{21} & b_{22} \end{pmatrix} \quad (2.14)$$

$$A \times B = C = \begin{pmatrix} c_{11} & c_{12} \\ c_{21} & c_{22} \end{pmatrix} \quad (2.15)$$

where:

$c_{11} = a_{11}.b_{11} + a_{12}.b_{21}$

$c_{12} = a_{11}.b_{12} + a_{12}.b_{22}$

$c_{21} = a_{21}.b_{11} + a_{22}.b_{21}$

$c_{22} = a_{21}.b_{12} + a_{22}.b_{22}$

For example, given matrices A and B, their multiplication is calculated as shown in matrix C as shown in the following:

$$A = \begin{pmatrix} 1 & 2 \\ 3 & 4 \end{pmatrix}$$

$$B = \begin{pmatrix} 5 & 6 \\ 7 & 8 \end{pmatrix}$$

$c_{11} = 1 \times 5 + 2 \times 7 = 5 + 14 = 19$

$c_{12} = 1 \times 6 + 2 \times 8 = 6 + 16 = 22$

$c_{21} = 3 \times 5 + 4 \times 7 = 15 + 28 = 43$

$c_{22} = 3 \times 6 + 4 \times 8 = 18 + 32 = 50$

Thus, $A \times B = C = \begin{pmatrix} c_{11} & c_{12} \\ c_{21} & c_{22} \end{pmatrix} = \begin{pmatrix} 19 & 22 \\ 43 & 50 \end{pmatrix}$

Scalar multiplication can also be applied to a matrix, where a scalar value is multiplied by each matrix element. Given matrix A in Equation (2.16) and a scalar value k, the result of multiplying matrix A by the scalar value k is as shown in Equation (2.17).

$$A = \begin{pmatrix} a & b & c \\ d & e & f \end{pmatrix} \tag{2.16}$$

$$A \times k = \begin{pmatrix} a \times k & b \times k & c \times k \\ d \times k & e \times k & f \times k \end{pmatrix} \tag{2.17}$$

For example, given matrix A and a scalar value $k = 3$, their product is calculated as follows:

$$A = \begin{pmatrix} 1 & 2 & 3 \\ 4 & 5 & 6 \end{pmatrix}$$

$$A \times k = \begin{pmatrix} 1 \times 3 & 2 \times 3 & 3 \times 3 \\ 4 \times 3 & 5 \times 3 & 6 \times 3 \end{pmatrix} = \begin{pmatrix} 3 & 6 & 9 \\ 12 & 15 & 18 \end{pmatrix}$$

2.1.3.4 Matrix transpose

Matrix transpose is an operation that produces a new matrix by flipping the rows and columns of a matrix. It involves creating a new matrix by changing the rows of a matrix into columns and its columns into rows. Given the matrix A in Equation (2.18), its transpose is denoted by A^T as shown in Equation (2.19).

$$A = \begin{pmatrix} a_{11} & a_{12} & a_{13} \\ a_{21} & a_{22} & a_{23} \\ a_{31} & a_{32} & a_{33} \end{pmatrix} \tag{2.18}$$

$$A^T = \begin{pmatrix} a_{11} & a_{21} & a_{31} \\ a_{12} & a_{22} & a_{32} \\ a_{13} & a_{23} & a_{33} \end{pmatrix} \tag{2.19}$$

For example, given matrix A, its transpose is as shown in the following:

$$A = \begin{pmatrix} 1 & 2 & 3 \\ 4 & 5 & 6 \\ 7 & 8 & 9 \end{pmatrix}$$

$$A^T = \begin{pmatrix} 1 & 4 & 7 \\ 2 & 5 & 8 \\ 3 & 6 & 9 \end{pmatrix}$$

2.1.3.5 Square and rectangular matrix

A square matrix is characterized by an equal number of rows (n) and the number of columns (m), denoted as $n = m$. It is differentiated from a rectangular matrix, where the number of rows and columns are not equal. Below is an example of a square matrix A, where $n = m = 3$ and a rectangular matrix B, where $n = 2$ and $m = 3$.

$$A = \begin{pmatrix} 1 & 2 & 3 \\ 4 & 5 & 6 \\ 7 & 8 & 9 \end{pmatrix}$$

$$B = \begin{pmatrix} 1 & 2 & 3 \\ 4 & 5 & 6 \end{pmatrix}$$

2.1.3.6 Triangular matrix

A triangular matrix is a special type of square matrix where all the elements above or below the diagonal are zeros. Depending on which side of the diagonal contains the non-zero elements, it can be classified either as an upper triangular matrix or a lower triangular matrix. As shown in the following, matrix A is an upper triangular matrix with non-zero elements located above the diagonal; matrix B is a lower triangular matrix with non-zero elements located below the diagonal.

$$A = \begin{pmatrix} 1 & 2 & 3 \\ 0 & 4 & 5 \\ 0 & 0 & 6 \end{pmatrix}$$

$$B = \begin{pmatrix} 1 & 0 & 0 \\ 4 & 5 & 0 \\ 0 & 0 & 6 \end{pmatrix}$$

2.1.3.7 Diagonal matrix

A diagonal matrix is a square matrix in which any value off the main diagonal is zero. Elements from top left to bottom right make up the primary diagonal. In the following example, the diagonal matrix is indicated by D.

$$D = \begin{pmatrix} 2 & 0 & 0 \\ 0 & 3 & 0 \\ 0 & 0 & 4 \end{pmatrix}$$

2.1.3.8 Identity matrix

An identity matrix is also a square matrix in which all elements along the diagonal are equal to 1, and all other elements off the diagonal are equal to zero. Matrix I is an example of the identity matrix.

$$I = \begin{pmatrix} 1 & 0 & 0 \\ 0 & 1 & 0 \\ 0 & 0 & 1 \end{pmatrix}$$

2.1.3.9 Matrix determinant

The determinant of a matrix is a scalar value that can be computed from the elements of a square matrix. It offers essential details about the matrix, such as whether it is invertible, singular, or neither. It is used in different fields of machine learning, data science, data mining, mathematics, and science, to mention a few, such as computing eigenvalues and eigenvectors, computing systems of linear equations, calculating areas and volumes, and analyzing transformations. The determinant of a square matrix A which is denoted by $\det(A)$ or $|A|$ can be evaluated differently depending on the dimension of a matrix. The formula for determinants of 2×2 and 3×3 matrices is given in Equations (2.20) and (2.21), respectively.

$$A = \begin{pmatrix} a_{11} & a_{12} \\ a_{21} & a_{22} \end{pmatrix}$$

$$\det(A) = a_{11} \times a_{22} - a_{21} \times a_{12} \tag{2.20}$$

$$A = \begin{pmatrix} a_{11} & a_{12} & a_{13} \\ a_{21} & a_{22} & a_{23} \\ a_{31} & a_{32} & a_{33} \end{pmatrix}$$

$$\det(A) = a_{11}(a_{22} \times a_{33} - a_{32} \times a_{23}) - a_{12}(a_{21} \times a_{33} - a_{31} \times a_{23}) \\ + a_{13}(a_{21} \times a_{32} - a_{31} \times a_{22}) \tag{2.21}$$

For larger matrices, the determinant can be calculated using different methods, such as cofactor expansion, LU decomposition, or Gaussian elimination, depending on the properties of the matrix and computational efficiency requirements.

As an example, given matrices $A_{2 \times 2}$ and $B_{3 \times 3}$ their determinants can be evaluated as shown in the following:

$$A = \begin{pmatrix} 2 & 3 \\ 1 & 4 \end{pmatrix}$$

$$\det(A) = 2 \times 4 - 1 \times 3 = 8 - 3 = 5$$

$$B = \begin{pmatrix} 1 & 2 & 3 \\ 0 & 1 & 4 \\ 5 & 6 & 0 \end{pmatrix}$$

$$\det(B) = 1(1 \times 0 - 4 \times 6) - 2(0 \times 0 - 4 \times 5) + 3(0 \times 6 - 1 \times 5)$$

$$\det(B) = 1(0 - 24) - 2(0 - 20) + 3(0 - 5)$$

$$\det(B) = 1(-24) - 2(-20) + 3(-5)$$

$$\det(B) = -24 + 40 - 15$$

$$\det(B) = 1$$

2.1.3.10 Adjugate of a matrix

The adjugate of a matrix, also known as the adjoint of the matrix (for matrix A, indicated by adj(A)), can be created in various methods depending on the matrix's dimension. In the case of a 2×2 matrix, Equation (2.22) specifies that the elements along the main diagonal are exchanged, and the signs of the elements off the main diagonal are modified. Conversely, for a 3×3 matrix, Equation (2.23) computes the cofactors (C_{ij}) of the matrix elements and proceeds to transpose the resulting matrix. Additionally, each element of the 3×3 adjugate matrix is the result of computing the determinant of the 2×2 sub-matrix obtained by removing the row and column of the corresponding element of the matrix multiplied by −1 if the sum of the row index and column index is odd, as shown in Equation (2.24).

$$A = \begin{pmatrix} a_{11} & a_{12} \\ a_{21} & a_{22} \end{pmatrix}$$

$$\text{adj}(A) = \begin{pmatrix} a_{22} & -a_{12} \\ -a_{21} & a_{11} \end{pmatrix} \qquad (2.22)$$

$$B = \begin{pmatrix} b_{11} & b_{12} & b_{13} \\ b_{21} & b_{22} & b_{23} \\ b_{31} & b_{32} & b_{33} \end{pmatrix}$$

$$\text{adj}(B) = \begin{pmatrix} C_{11} & C_{12} & C_{13} \\ C_{21} & C_{22} & C_{23} \\ C_{31} & C_{32} & C_{33} \end{pmatrix}^T = \begin{pmatrix} C_{11} & C_{21} & C_{31} \\ C_{12} & C_{22} & C_{32} \\ C_{13} & C_{23} & C_{33} \end{pmatrix} \qquad (2.23)$$

where:

$$C_{ij} = (-1)^{i+j} \times \det(M_{ij}) \qquad (2.24)$$

where:

M_{ij} is the resulting 2×2 sub-matrix after removing the i-th row and j-th column. For example, given matrices $A_{2\times 2}$ and $B_{3\times 3}$ their adjugates are calculated as follows:

$$A = \begin{pmatrix} 2 & 3 \\ 1 & 4 \end{pmatrix}$$

$$\text{adj}(A) = \begin{pmatrix} 4 & -3 \\ -1 & 2 \end{pmatrix}$$

$$B = \begin{pmatrix} 1 & 2 & 3 \\ 0 & 1 & 4 \\ 5 & 6 & 0 \end{pmatrix}$$

$$M_{11} = \det\begin{pmatrix} 1 & 4 \\ 6 & 0 \end{pmatrix} = (1 \times 0) - (4 \times 6) = -24$$

Since $i + j = 1 + 1 = 2, (-1)^{i+j} = (-1)^2 = 1$, thus, $C_{11} = (1) \times (-24) = -24$

$$M_{12} = \det\begin{pmatrix} 0 & 4 \\ 5 & 0 \end{pmatrix} = (0 \times 0) - (4 \times 5) = -20$$

Since $i + j = 1 + 2 = 3, (-1)^{i+j} = (-1)^3 = -1$, thus, $C_{12} = (-1) \times (-20) = 20$

$$M_{13} = \det\begin{pmatrix} 0 & 1 \\ 5 & 6 \end{pmatrix} = (0 \times 6) - (1 \times 5) = -5$$

Since $i + j = 1 + 3 = 4, (-1)^{i+j} = (-1)^4 = 1$, thus, $C_{13} = (1) \times (-5) = -5$

$$M_{21} = \det\begin{pmatrix} 2 & 3 \\ 6 & 0 \end{pmatrix} = (2 \times 0) - (3 \times 6) = -18$$

Since $i + j = 2 + 1 = 3, (-1)^{i+j} = (-1)^3 = -1$, thus, $C_{21} = (-1) \times (-18) = 18$

$$M_{22} = \det\begin{pmatrix} 1 & 3 \\ 5 & 0 \end{pmatrix} = (1 \times 0) - (3 \times 5) = -15$$

Since $i + j = 2 + 2 = 4, (-1)^{i+j} = (-1)^4 = 1$, thus, $C_{22} = (1) \times (-15) = -15$

$$M_{23} = \det\begin{pmatrix} 1 & 2 \\ 5 & 6 \end{pmatrix} = (1 \times 6) - (2 \times 5) = -4$$

Since $i + j = 2 + 3 = 5, (-1)^{i+j} = (-1)^5 = -1$, thus, $C_{23} = (-1) \times (-4) = 4$

$$M_{31} = \det\begin{pmatrix} 2 & 3 \\ 1 & 4 \end{pmatrix} = (2 \times 4) - (1 \times 3) = 5$$

Since $i + j = 3 + 1 = 4, (-1)^{i+j} = (-1)^4 = 1$, thus, $C_{31} = (1) \times (5) = 5$

$$M_{32} = \det\begin{pmatrix} 1 & 3 \\ 0 & 4 \end{pmatrix} = (1 \times 4) - (3 \times 0) = 4$$

Since $i + j = 3 + 2 = 5, (-1)^{i+j} = (-1)^5 = -1$, thus, $C_{32} = (-1) \times (4) = -4$

$$M_{33} = \det\begin{pmatrix} 1 & 2 \\ 0 & 1 \end{pmatrix} = (1 \times 1) - (2 \times 0) = 1$$

Since $i + j = 3 + 3 = 6, (-1)^{i+j} = (-1)^6 = 1$, thus, $C_{32} = (1) \times (1) = 1$

$$C = \begin{pmatrix} -24 & 20 & -5 \\ 18 & -15 & 4 \\ 5 & -4 & 1 \end{pmatrix} \text{ and adj}(B) = \begin{pmatrix} -24 & 18 & 5 \\ 20 & -15 & -4 \\ -5 & 4 & 1 \end{pmatrix}$$

2.1.3.11 Singular and non-singular matrix

A singular matrix and a non-singular matrix are characterized by having determinants of zero and non-zero values, respectively. Consequently, the inverse of a singular matrix does not exist, whereas the inverse of a non-singular matrix exists. For example, given a singular matrix A, there is no matrix A^{-1}, such that $A \times A^{-1}$ or $A^{-1} \times A = I$, where I is an identity matrix. However, if a matrix A is non-singular, there exists A^{-1} such that $A \times A^{-1}$ or $A^{-1} \times A = I$.

2.1.3.12 Matrix inversion

Matrix inversion is a process of finding the inverse for a square non-singular matrix. Given matrix A, its inverse is denoted by A^{-1}. The inverse of a matrix is computed by dividing each element of the adjugate by the determinant of the matrix. The formula for computing the inverse of a matrix A is given in Equation (2.25).

$$A = \begin{pmatrix} a_{11} & a_{12} & a_{13} \\ a_{21} & a_{22} & a_{23} \\ a_{31} & a_{32} & a_{33} \end{pmatrix}$$

$$A^{-1} = \frac{1}{\det(A)} \times \text{adj}(A) = \frac{1}{\det(A)} \times \begin{pmatrix} C_{11} & C_{21} & C_{31} \\ C_{12} & C_{22} & C_{32} \\ C_{13} & C_{23} & C_{33} \end{pmatrix} \quad (2.25)$$

For example, given matrix B, its inverse matrix is calculated as shown in the following:

$$B = \begin{pmatrix} 1 & 2 & 3 \\ 0 & 1 & 4 \\ 5 & 6 & 0 \end{pmatrix}$$

$$\det(B) = 1(1\times 0 - 4\times 6) - 2(0\times 0 - 4\times 5) + 3(0\times 6 - 1\times 5) = 1$$
$$\neq 0 \text{ (i.e., } B \text{ is a non-singular matrix)}$$

$$\text{adj}(B) = \begin{pmatrix} -24 & 18 & 5 \\ 20 & -15 & -4 \\ -5 & 4 & 1 \end{pmatrix}$$

$$A^{-1} = \frac{1}{1} \times \begin{pmatrix} -24 & 18 & 5 \\ 20 & -15 & -4 \\ -5 & 4 & 1 \end{pmatrix} = \begin{pmatrix} -24 & 18 & 5 \\ 20 & -15 & -4 \\ -5 & 4 & 1 \end{pmatrix}$$

2.1.3.13 Eigenvectors and eigenvalues

An eigenvector is a non-zero vector v that changes in magnitude but retains its direction when a square matrix is applied to it as a linear transformation (i.e., when multiplied by an eigenvalue). The eigenvalue is a scalar value that represents the scaling factor of the eigenvector and indicates the extent to which the eigenvector has been stretched. Eigenvectors and eigenvalues are used to identify directions and patterns in data, reduce complexity, and make sense of information. Mathematically, given a square matrix A, the relationship of an eigenvector v of matrix A and its corresponding eigenvalue λ is shown in Equation (2.26). In addition, given a matrix A, the eigenvalue can be computed using Equation (2.27), whereas the eigenvector can be computed using Equation (2.28).

$$Av = \lambda v \tag{2.26}$$

$$\det(A - \lambda I) = 0 \tag{2.27}$$

$$(A - \lambda I)v = 0 \tag{2.28}$$

For example, given matrix A, its eigenvalues and eigenvectors are calculated as follows:

$$A = \begin{pmatrix} 2 & 1 \\ 1 & 3 \end{pmatrix}$$

$$\det\left(\begin{pmatrix} 2 & 1 \\ 1 & 3 \end{pmatrix} - \lambda \begin{pmatrix} 1 & 0 \\ 0 & 1 \end{pmatrix}\right) = 0$$

$$\det\left(\begin{pmatrix} 2 & 1 \\ 1 & 3 \end{pmatrix} - \begin{pmatrix} \lambda & 0 \\ 0 & \lambda \end{pmatrix}\right) = 0$$

$$\det\left(\begin{pmatrix} 2-\lambda & 1 \\ 1 & 3-\lambda \end{pmatrix}\right) = 0$$

$$\det\begin{pmatrix} 2-\lambda & 1 \\ 1 & 3-\lambda \end{pmatrix} = 0$$

$$((2-\lambda)(3-\lambda)) - (1)(1) = (6 - 5\lambda + \lambda^2) - 1 = \lambda^2 - 5\lambda + 5 = 0$$

Finding the eigenvalue λ by using the quadratic formula:

$$\lambda = \frac{-b \pm \sqrt{b^2 - 4ac}}{2a}$$

$$\lambda = \frac{-(-5) \pm \sqrt{5^2 - (4 \times 1 \times 5)}}{2 \times 1} = \frac{5 \pm \sqrt{25 - 20}}{2} = \frac{5 \pm \sqrt{5}}{2}$$

$$\lambda_1 = \frac{5 + \sqrt{5}}{2} \text{ and } \lambda_2 = \frac{5 - \sqrt{5}}{2}$$

Using Equation (2.28), the eigenvector for the eigenvalue $\lambda_1 = \frac{5 + \sqrt{5}}{2}$ can be computed as follows:

$$(A - \lambda_1 I) v = \left(\begin{pmatrix} 2 & 1 \\ 1 & 3 \end{pmatrix} - \frac{5 + \sqrt{5}}{2} \begin{pmatrix} 1 & 0 \\ 0 & 1 \end{pmatrix} \right) \begin{pmatrix} v_1 \\ v_2 \end{pmatrix} = 0$$

$$\left(\begin{pmatrix} 2 & 1 \\ 1 & 3 \end{pmatrix} - \begin{pmatrix} \frac{5 + \sqrt{5}}{2} & 0 \\ 0 & \frac{5 + \sqrt{5}}{2} \end{pmatrix} \right) \begin{pmatrix} v_1 \\ v_2 \end{pmatrix} = 0$$

$$\begin{pmatrix} 2 - \frac{5 + \sqrt{5}}{2} & 1 \\ 1 & 3 - \frac{5 + \sqrt{5}}{2} \end{pmatrix} \begin{pmatrix} v_1 \\ v_2 \end{pmatrix} = 0$$

$$\begin{pmatrix} \frac{-1 + \sqrt{5}}{2} & 1 \\ 1 & \frac{1 + \sqrt{5}}{2} \end{pmatrix} \begin{pmatrix} v_1 \\ v_2 \end{pmatrix} = 0$$

$$\frac{-1 + \sqrt{5}}{2} v_1 + v_2 = 0 \tag{2a}$$

$$v_1 + \frac{1 + \sqrt{5}}{2} v_2 = 0 \tag{2b}$$

From Equation (2a):

$$v_2 = \frac{1-\sqrt{5}}{2} v_1 \tag{2c}$$

Substituting Equation (2c) in Equation (2b):

$$v_1 + \left(\frac{1+\sqrt{5}}{2}\right)\left(\frac{1-\sqrt{5}}{2} v_1\right) = 0$$

$$v_1 + \frac{1-5}{4} v_1 = 0$$

$$v_1 + \frac{-4}{4} v_1 = 0$$

$$v_1 - v_1 = 0$$

$$0 = 0$$

This equation is always true, which means there are infinitely many solutions for v_1, any non-zero value can be chosen for v_1 and the corresponding value of v_2 can be computed by using Equation (2c). Suppose $v_1 = 1$, then the value of v_2 can be obtained as follows:

$$v_2 = \frac{1-\sqrt{5}}{2} v_1 = \frac{1-\sqrt{5}}{2} \times 1 = \frac{1-\sqrt{5}}{2}$$

Therefore, the possible value of the eigenvector with its corresponding eigenvalue $\lambda_1 = \frac{5+\sqrt{5}}{2}$ is $\begin{pmatrix} 1 \\ \frac{1-\sqrt{5}}{2} \end{pmatrix}$.

In the same fashion, the possible value of the eigenvector for the eigenvalue $\lambda_2 = \frac{5-\sqrt{5}}{2}$ can be calculated.

2.2 STATISTICS CONCEPTS

In machine learning, statistics is the application of statistical concepts and methods to data analysis, prediction, and model performance assessment. Effective model training and interpretation is made possible by its foundation in the understanding of uncertainty, variability, and linkages within datasets.

2.2.1 Use of statistics in machine learning

In every aspect of machine learning, statistics plays an important role in algorithm selections, developments, and real-world applications. The core of the machine learning process is formed by practitioners' ability to understand, assess, and extract insights from data. Machine learning workflow begins with data preprocessing tasks such as data cleaning, normalization, and modeling. However, advanced modeling techniques such as regression, classification, and clustering statistics inform the entire spectrum of machine learning processes. Furthermore, through statistical techniques, outliers and extreme values are detected; missing values are substituted and normalized, therefore guaranteeing the accuracy and dependability of the dataset.

Descriptive statistics and visualization techniques assist in exploring data characteristics and relationships, thereby guiding feature selection and dimensionality reduction in datasets. Hence, statistical methods are used to build models for both supervised and unsupervised learning tasks. Additionally, metrics are used to measure how well models perform and generalize.

Statistics improves decision-making in complicated settings by enabling probabilistic modeling and uncertainty quantification. However, frameworks for representing and arguing about uncertainty are provided by methods such as Bayesian inference and probabilistic graphical models. Additionally, statistics guides feature selection and engineering efforts, identifying informative features and reducing dimensionality while preserving essential data structure. Furthermore, its holistic integration across the machine learning pipeline empowers practitioners to unlock the potential of data across various domains. Statistics is basically the base on which machine learning grows and develops. It helps professionals find useful insights and make informed decisions in a world that is becoming more and more data-driven.

2.2.2 Types of statistics

Statistics can be broken down into two types: descriptive statistics and inferential statistics.

2.2.2.1 Descriptive statistics

Descriptive statistics is the study of how to organize, summarize, and show data in a way that makes sense and gives us useful information. Its goal is to show the most important features of a dataset, including trends, ranges, and patterns of distribution. At the cutting edge of data analysis, descriptive statistics tries to turn complicated datasets into concepts that are easy to understand and comprehend. Additionally, it is important to note that descriptive statistics do not draw conclusions about the whole community or anything bigger than the dataset it is looking at. Instead, it shows and summarizes the dataset's natural properties.

    i. **Measures of Central Tendency**
- **Mean**
  Mean is the most commonly used measure of central tendency. It is computed by adding up all the values of the elements in the list and then dividing that number by the number of elements. Equation (2.29) illustrates the computation of the mean.

$$\text{Mean} = \frac{\text{Sum of elements}}{\text{Total number of elements}} \qquad (2.29)$$

Consider a class whose students have obtained the following marks out of 100: 45, 55, 60, 75, 80, 55, 37, 39, 25, 48, 37, and 68. The mean is calculated as shown in the following:
Sum of elements = 45 + 55 + 60 + 75 + 80 + 55 + 37 + 39 + 25 + 48 + 37 + 68 = 624
Total number of elements = 12.
Thus,

$$Mean = \frac{624}{12} = 52$$

- **Median**

The median of a set of numbers is the middle value when the numbers are arranged in ascending or descending order. If the set contains an odd number of values, the median is the middle number. If the set contains an even number of values, the median is the average of the two middle numbers. This measure is less sensitive to extreme values (i.e., outliers) compared to the mean.

Consider the same example of a class whose students have obtained the following marks out of 100: 45, 55, 60, 75, 80, 55, 37, 39, 25, 48, 37, and 68. The median is calculated as shown in the following.
First, arrange the values in ascending order:
25, 37, 37, 39, 45, **48, 55**, 55, 60, 68, 75, 80
Since the number of elements is 12 (i.e., even), the median value will be the average of the sixth (i.e., 48) and seventh (i.e., 55) elements.
Thus,

$$Median = \frac{48 + 55}{2} = \frac{103}{2} = 51.5$$

- **Mode**

The mode is the value that appears most frequently in a set of data. The set of data may have one mode, more than one mode (i.e., multimodal), or no mode at all (i.e., when all values occur with the same frequency). The mode is useful in filling the missing values for categorical data. Consider a class whose students have obtained the following marks out of 100: 45, 55, 60, 75, 80, 55, 37, 39, 25, 48, 37, and 68. The mode is calculated as shown in the following. For simplicity in identifying the mode, it is advised to arrange the values in ascending order as follows:

25, **37, 37,** 39, 45, 48, 55, 55, 60, 68, 75, 80

Since 37 appears most frequently (i.e., 2 times compared to others) in the set of data, then the mode is 37.

ii. **Measures of Dispersion**

Measures of dispersion provide insights into the variability of data from the central tendency such as mean or median. They provide valuable insights into

how widely the values are spread from the center of the distribution, helping to understand the distribution and potential outliers within the dataset.
- **Range**

The range is a measure that indicates the extent of variation within a dataset by quantifying the difference between the largest and smallest values. It is calculated by subtracting the minimum value from the maximum value. For example, in a dataset of test scores {65, 72, 80, 85, 92}, the range would be **92** (i.e., the largest value) minus **65** (i.e., the smallest value), resulting in a range of **27**.
- **Percentiles**

A percentile is a statistical measure that signifies the value below which a specific percentage of observations in a dataset lies. For instance, the 20th percentile denotes that the value falls below 20% of the dataset. For example, if the 20th percentile score is 35, it means that 20% of the total observations have a value less than 35. Consider a dataset showing the heights (in inches) of ten individuals: {66, 75, 64, 65, 67, 72, 68, 70, 62, and 60}. The percentiles are calculated as shown in the following steps.

1. **Sort the Data**: Arrange the data in ascending order.

   {60, 62, 64, 65, 66, 67, 68, 70, 72, 75}

2. **Calculate the Position**: Determine the position of the desired percentile in the dataset using the formula given in Equation (2.30).

$$\text{Position} = \left(\frac{P}{100}\right) \times (n+1) \tag{2.30}$$

where:

P is the desired percentile (e.g., 25th percentile, 50th percentile, etc.), and
n is the total number of data points in the dataset.

For instance, the 25th, 50th, and 90th percentile positions are computed as follows:

25th Percentile:

$$\text{Position} = \left(\frac{25}{100}\right) \times (10+1) = 0.25 \times 11 = 2.75^{\text{th}} \text{ value.}$$

50th Percentile:

$$\text{Position} = \left(\frac{50}{100}\right) \times (10+1) = 0.5 \times 11 = 5.5^{\text{th}} \text{ value.}$$

90th Percentile:

$$\text{Position} = \left(\frac{90}{100}\right) \times (10+1) = 0.9 \times 11 = 9.9^{\text{th}} \text{ value.}$$

3. **Interpolate if necessary**: If the position is an integer, the percentile is identified as the value at that position. However, if the position is not an integer, interpolate between the values at the nearest lower and higher positions to find the exact value of the percentile. Consider the integer portion as R (i.e., the number to the left of the decimal point) and the fractional portion as FR (i.e., the number to the right of the decimal point), the value at the nearest lower position as L and the value at the nearest higher position as H, the percentile can be computed using Equation (2.31).

$$\text{Percentile value} = FR(H-L) + L \tag{2.31}$$

For the 25th percentile in a dataset with 10 data points is the 2.75th value, making $R = 2$, $FR = 0.75$, $H = 64$, and $L = 62$, the 25th percentile is computed as shown in the following:

$$25^{th} \text{ Percentile} = 0.75(64-62) + 62 = 0.75(2) + 62 = 63.5$$

For the 50th percentile in a dataset with 10 data points is the 5.5th value, making $R = 5$, $FR = 0.5$, $H = 67$, and $L = 66$, the 50th percentile is computed as shown in the following:

$$50^{th} \text{ Percentile} = 0.5(67-66) + 66 = 0.5(1) + 66 = 66.5$$

For the 90th percentile in a dataset with 10 data points is the 9.9th value, making $R = 9$, $FR = 0.9$, $H = 75$, and $L = 72$, the 90th percentile is computed as shown in the following:

$$90^{th} \text{ Percentile} = 0.9(75-72) + 72 = 0.9(3) + 72 = 74.7$$

- **Quartiles**

Quartiles are measures that divide a dataset into four equal parts, each containing approximately 25% of the data. In terms of percentiles, the first quartile (i.e., Q1) corresponds to the 25th percentile, the second quartile (i.e., Q2) corresponds to the 50th percentile (i.e., median), and the third quartile (i.e., Q3) corresponds to the 75th percentile. In order to calculate Q1, Q2, and Q3, refer to the approach used to compute the percentiles.

- **Interquartile Range**

In descriptive statistics, the interquartile range (IQR) is a measure of statistical spread or dispersion. It is expressed mathematically as the difference between the first (i.e., 25th percentile or Q1) and third (i.e., 75th percentile or Q3) quartiles of the data. Figure 2.1 illustrates a box plot, which shows minimum value, maximum value, IQR, lower quartile, and upper quartile. The box plot is used to identify and handle outliers and extreme values in the datasets.

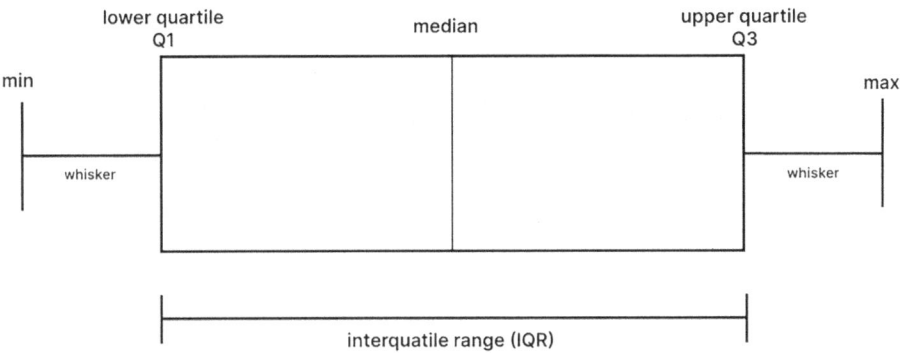

FIGURE 2.1 IQR using a box plot.

Equation (2.32) illustrates how to compute IQR by subtracting the first quartile (Q1) from the third quartile (Q3).

$$IQR = Q3 - Q1 \tag{2.32}$$

where:

$Q1$ is the first quartile (25th percentile).
$Q3$ is the third quartile (75th percentile).

For example, given $Q1 = 6$ and $Q3 = 13.5$, then IQR = $13.5 - 6 = 7.5$

- **Mean Absolute Deviation**
  Mean Absolute Deviation (MAD) is a statistical measure that describes the variability in a dataset. It measures how much of an average absolute difference each data point has from the dataset average. Furthermore, MAD offers a reliable and understandable measure of variability in a dataset. The computation of MAD involves calculating the *mean* of all the data points, finding the absolute difference between each data point and the mean, summing up all the absolute differences and dividing the sum by the total number of data points, n, as shown in Equation (2.33).

$$MAD = \frac{\sum_{i=1}^{n} |x_i - \text{mean}|}{n} \tag{2.33}$$

where:

mean is the mean of the dataset.
n is the total number of data points.
x_i represents each data point in the dataset.

For example, the MAD for the dataset {3, 7, 8, 5, 12, 14, 21, 13, and 18} can be calculated using Equation (2.33) as shown below:

$$\text{mean} = \frac{3+7+8+5+12+14+21+13+18}{9} = \frac{101}{9} = 11.2$$

$$\text{sum of absolute differences} = |3-11.2| + |7-11.2| + |8-11.2| + |5-11.2| + |12-11.2|$$
$$+ |14-11.2| + |21-11.2| + |13-11.2| + |18-11.2|$$

$$\text{sum of absolute differences} = 8.2 + 4.2 + 3.2 + 6.2 + 0.8 + 2.8 + 9.8 + 1.8 + 6.8 = 43.8$$

$$\text{MAD} = \frac{43.8}{9} = 4.87$$

- **Variance**

Variance measures the dispersion of data points from the mean. It is calculated by finding the mean of all the data points, summing the squared differences between the data points and the mean, and then dividing the sum by the total number of data points, as shown in Equation (2.34). As opposed to MAD, variance uses the squares of differences between the data points and the mean. The challenge with variance is in its unit inconsistency due to squaring, which makes it less intuitive for interpretation. Consequently, the standard deviation is often preferred, as it provides a measure of dispersion in the same units as the original data. Variance is computed as shown in Equation (2.34).

$$\text{Variance} = \frac{\sum_{i=1}^{n}(x_i - \text{mean})^2}{n} \qquad (2.34)$$

For example, the variance for the dataset {3, 7, 8, 5, 12, 14, 21, 13, and 18} can be calculated as follows using Equation (2.34).

$$\text{mean} = \frac{3+7+8+5+12+14+21+13+18}{9} = \frac{101}{9} = 11.2$$

The sum of the squared differences between each data point and the mean:

$$\text{sum} = (3-11.2)^2 + (7-11.2)^2 + (8-11.2)^2 + (5-11.2)^2 + (12-11.2)^2$$
$$+ (14-11.2)^2 + (21-11.2)^2 + (13-11.2)^2 + (18-11.2)^2$$

$$\text{sum} = (8.2)^2 + (4.2)^2 + (3.2)^2 + (6.2)^2 + (0.8)^2 + (2.8)^2 + (9.8)^2 + (1.8)^2 + (6.8)^2$$

$$\text{sum} = 67.24 + 17.64 + 10.24 + 38.44 + 0.64 + 7.84 + 96.04 + 3.24 + 46.24 = 287.56$$

$$\text{Variance} = \frac{287.56}{9} = 31.95$$

- **Standard Deviation**

Standard deviation is a measure of dispersion of data points from the mean. It can be calculated as the square root of the variance, as shown in Equation (2.35). Therefore, it provides a measure of variability in the same units as the original data. Additionally, for a data point, a higher standard deviation value indicates greater dispersion from the mean, whereas a lower standard deviation value suggests closer proximity to the mean. The standard deviation is computed as shown in Equation (2.35).

$$\text{Standard Deviation} = \sqrt{\frac{\sum_{i-1}^{n}(x_i - \text{mean})^2}{n}} \quad (2.35)$$

For example, the standard deviation for the dataset {3, 7, 8, 5, 12, 14, 21, 13, and 18} can be calculated using Equation (2.35) as follows:
Since the variance of the data is 31.95, then:

standard deviation = $\sqrt{31.95}$ = 5.65

- **Median Absolute Deviation**

The Median Absolute Deviation (MedAD) is a robust measure of the variability or dispersion of a dataset. It is calculated as the median of the absolute differences between each data point and the median of the dataset. However, MedAD is less sensitive to outliers and extreme values compared to standard deviation, making it suitable for evaluating datasets.

2.2.2.2 Inferential statistics

Inferential statistics is the process of making inferences about a broader population from a sample of data that has been taken from that population. Additionally, insights are gained and predictions that apply to the full population are established through statistical testing and analysis samples. Regression analysis, hypothesis testing, data manipulation, and visualization are some of the approaches employed. This process assists in identifying patterns and extracts valuable information. Even in cases where data availability is restricted, inferential statistics can be used to draw defensible inferences about populations and make well-informed decisions.

2.2.3 Types of data

Data can be broadly categorized into two types, numerical and categorical as described in the following subsections.

2.2.3.1 Numerical data

Numerical data consists of quantifiable data, such as height, weight, temperature, or test results. Additionally, numerical data refers to quantities expressed as integers or decimal numbers, often known as floating-point numbers. This data can be classified into two primary categories:

(i) Discrete numerical data are precise and separate quantities, typically indicating counts or categories. Examples include the rank of students in a classroom or the number of faculties in a department.
(ii) Continuous numerical values include values that can take on any real number within a certain range. Unlike discrete values, continuous values have an infinite number of possible values. An example is the salary of an employee, which can vary continuously within a certain range.

2.2.3.2 Categorical data

Categorical data represents qualitative values that are typically divided into categories or groups. It is often expressed as strings or characters. Examples include names, colors, or any type of non-numeric labels. This type of data is commonly categorized into two main types: ordinal and nominal. Ordinal categorical values can be meaningfully ranked or ordered, but the intervals between rankings may not be uniform. Examples include student grades (e.g., A, B, C) and satisfaction ratings (e.g., high, medium, low). Nominal categorical values can be represented in various groups or names, with no intrinsic order or ranking. They are composed of distinct categories with no implicit hierarchy. Examples include colors (red, blue, green), courses (math, science, history), and fruit varieties (apple, banana, orange).

2.2.4 Data distribution

Data distribution refers to how a set of data is spread out and dispersed throughout a range of possible values. It can be graphically represented using a histogram, frequency polygon, or box plot. Understanding data distribution is crucial because it shows patterns that are not immediately obvious when looking at the data itself. Data distribution can reveal if the data is symmetrical, how densely the data is clustered, and whether the data is skewed.

2.2.4.1 Normal distribution in statistics

Normal distribution is a type of data distribution that is also known as a Gaussian distribution. It is defined by its mean and standard deviation and is characterized by a bell-shaped curve, as shown in Figure 2.2. Normal distribution is prevalent in many datasets used in machine learning. For datasets that do not naturally follow this distribution, efforts are often made to transform the data into a normal distribution due to its favorable properties. Additionally, many machine learning algorithms perform optimally on data that approximates a normal distribution as the distribution mirrors real-world phenomena, such as salary distributions, where the majority of employees fall within a medium range, with fewer at the extremes of low or high salaries. The normal distribution aligns with the Empirical Rule, which states that about 68% of the data falls

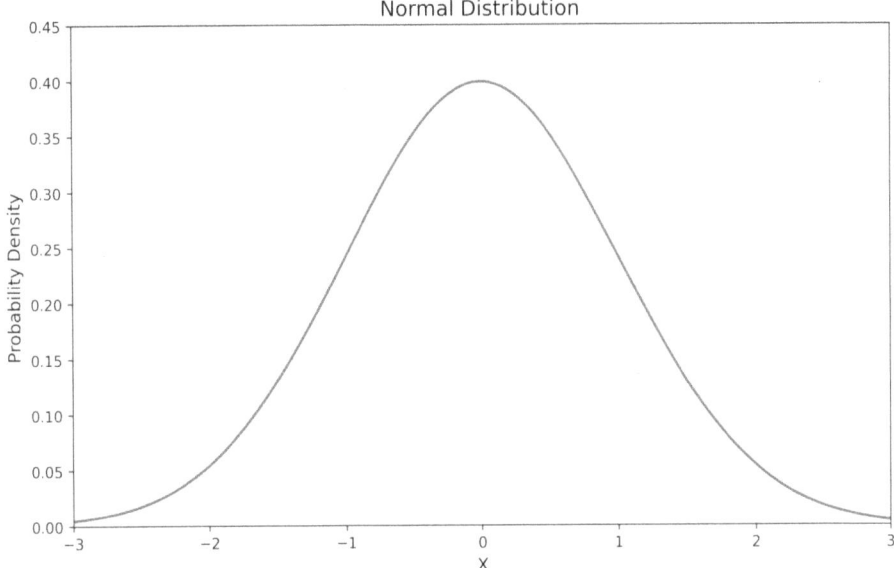

FIGURE 2.2 Normal distribution curve.

within one standard deviation of the mean, 95% falls within two standard deviations, and 99.7% falls within three standard deviations.

The normal distribution aligns with the Empirical Rule. The rule outlines the proportion of data falling within specific ranges of standard deviations from the mean. According to the rule, approximately 68% of the data lies within one standard deviation of the mean. This means that the majority of observations in a normally distributed dataset are clustered within a relatively narrow range around the mean, as shown in Figure 2.3.

The Empirical Rule also states that about 95% of the data falls within two standard deviations of the mean. This wider interval encompasses a significant portion of the dataset, indicating a broader dispersion of observations from the mean, as shown in Figure 2.4. While there is greater variability within this range compared to the first standard deviation, the majority of data points still exhibit a pattern consistent with the normal distribution.

The Empirical Rule also asserts that nearly 99.7% of the data falls within three standard deviations of the mean. This extensive range covers the vast majority of observations in a normally distributed dataset, reflecting the symmetrical nature of the bell-shaped curve, as shown in Figure 2.5. The diminishing proportion of data beyond three standard deviations underscores the rare occurrence of extreme values in a dataset following the normal distribution.

2.2.4.2 Skewness

Skewness measures the asymmetry of distribution as depicted in histograms or Kernel Density Estimation (KDE) plots and is usually characterized by a pronounced peak toward the mode of the data. Skewness is commonly categorized into two

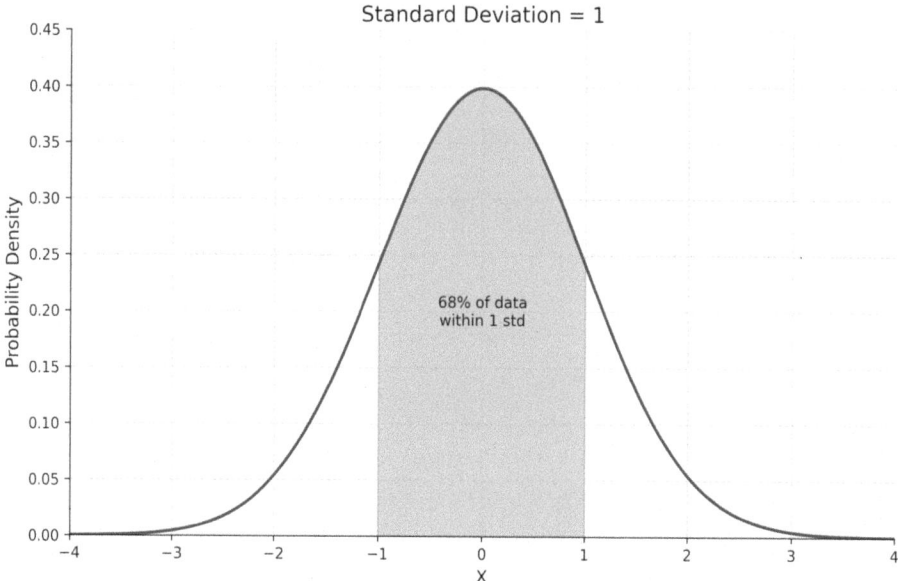

FIGURE 2.3 68% of all values are within 1 standard deviation of the mean value.

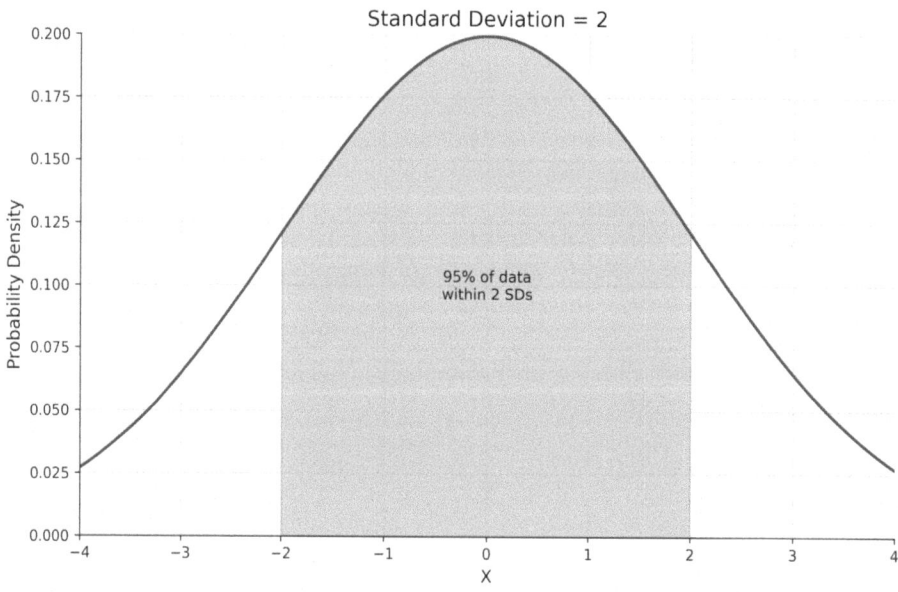

FIGURE 2.4 95% of all values are within 2 standard deviation of mean value.

types: left-skewed (i.e., negative skewness) and right-skewed (i.e., positive skewness) as shown in Figure 2.6. Additionally, some consider a third category: symmetric distribution, which is indicative of a normal distribution. A right-skewed distribution is characterized by a long tail extending toward the positive axis. A suitable

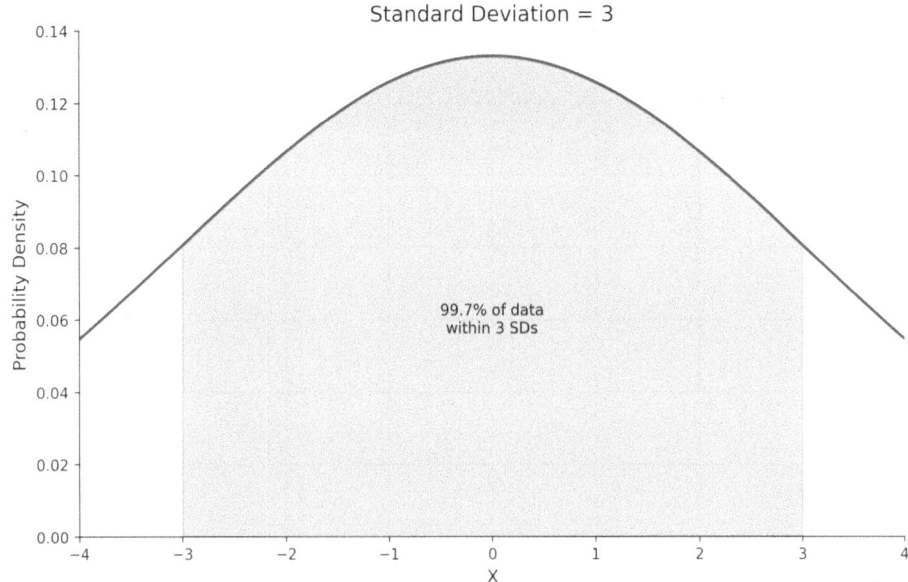

FIGURE 2.5 99.7% of all values are within 3 standard deviation of mean value.

example of right-skewed data is wealth distribution, where only a small percentage of individuals possess very high wealth, while the majority falls within the middle range. On the other hand, a left-skewed distribution is marked by a long tail extending toward the negative axis. For instance, consider the distribution of grades among students, where fewer students receive lower grades, while the majority of them fall within the passing category.

2.2.4.3 Central limit theorem

The Central Limit Theorem (CLT) states that "regardless of population distribution, the sampling distribution of the sample mean approaches a normal distribution as sample size increases." Figure 2.7 illustrates the theorem, which allows machine learning practitioners to draw conclusions about population parameters based on sample means even when the population distribution is unknown.

2.2.5 Applied statistical inference

This section looks into the practical applications of inferential statistics. The concept entails drawing conclusions about a population using sample data. Applied statistical inference comprises utilizing statistical methods to assess data, derive meaningful insights, and drive decision-making across a variety of domains. As a result, this part focuses on the application of linear regression as a fundamental approach for predictive modeling and statistical analysis.

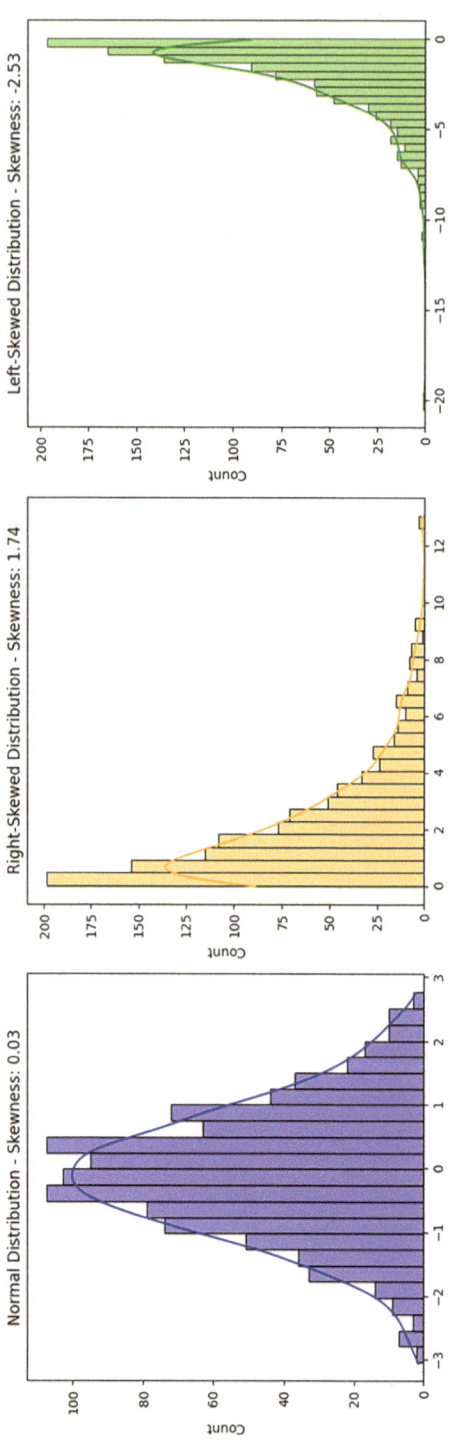

FIGURE 2.6 Skewness.

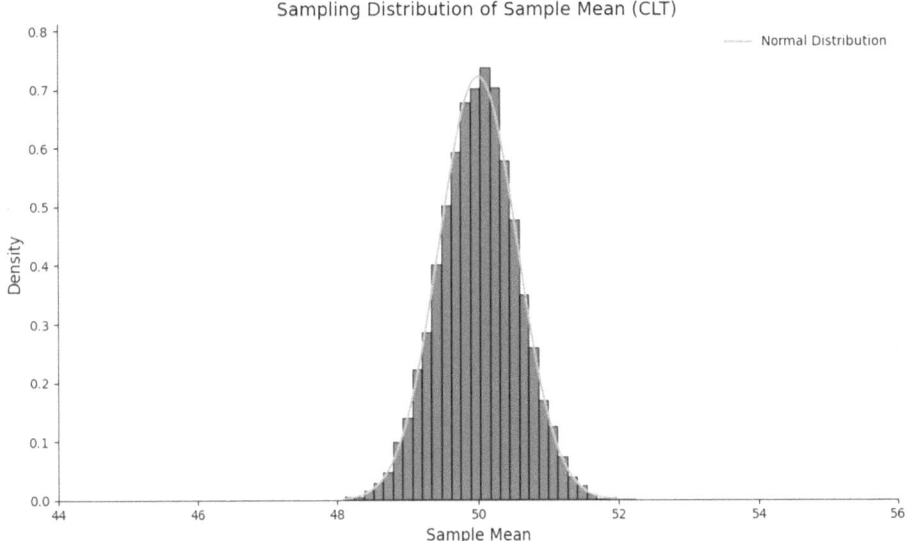

FIGURE 2.7 Demonstration of the Central Limit Theorem.

2.2.5.1 Linear regression

Linear regression is a statistical technique that models the relationship between a dependent variable and independent variables by fitting the regression line to observed data. As a result, the relationship between the variables is thought to be linear, meaning that changes in the independent variable produce changes in the dependent variable(s) at the same pace. Additionally, it is one of the most basic and widely applied approaches in statistical modeling and predictive analysis. However, the objective of linear regression is to find the best-fitting line (or plane, in the case of numerous independent variables) that minimizes the difference between the observed data points and the predicted values provided by the linear equation. Hence, this line is then used to forecast the dependent variable using the values of the independent variables. This section covers two types of linear regression: univariate and multivariate linear regressions.

   i. **Univariate linear regression**
   Univariate linear regression or simple linear regression describes the relationship between a single independent variable (X) and one dependent variable (Y). Hence, the univariate linear regression model is described by Equation (2.36).

$$Y = \beta_0 + \beta_1 X + \varepsilon \quad (2.36)$$

where:

β_0 is the intercept, representing the value of Y when X is zero.
β_1 is the slope, representing the rate of change in Y for a one-unit change in X.
ε is the error term, representing the difference between the observed and predicted values of Y ($\varepsilon = Y_i - \hat{Y}_i$).

The goal of univariate linear regression is to estimate the values of β_0 and β_1 that minimize the sum of squared differences between the observed and predicted values of Y, typically using the method of least squares. The estimation of coefficients can be obtained through the following steps.

    a. **Calculate the Mean**: Compute the means of the dependent variable Y and the independent variable X as shown in Equations (2.37) and (2.38), respectively.

$$\bar{Y} = \frac{1}{n}\sum_{i=1}^{n} Y_i \qquad (2.37)$$

$$\bar{X} = \frac{1}{n}\sum_{i=1}^{n} X_i \qquad (2.38)$$

where:

n is the total number of samples.

    b. **Calculate Covariance and Variance**: Compute the sample covariance between X and Y and the sample variance of X as shown in Equation (2.39) and (2.40), respectively.

$$\text{cov}(X,Y) = S_{XY} = \frac{1}{n-1}\sum_{i=1}^{n}(X_i - \bar{X})(Y_i - \bar{Y}) \qquad (2.39)$$

$$\text{Var}(X) = S_{XX} = \frac{1}{n-1}\sum_{i=1}^{n}(X_i - \bar{X})^2 \qquad (2.40)$$

    c. **Estimate Slope**: To estimate the slope (β_1) of the regression line, the covariance of X and Y is divided by the variance of X, as shown in Equation (2.41).

$$\hat{\beta}_1 = \frac{\text{cov}(X,Y)}{\text{Var}(X)} = \frac{S_{XY}}{S_{XX}} \qquad (2.41)$$

    d. **Estimate Intercept**: To estimate the intercept (β_0) of the regression line, the slope ($\hat{\beta}_1$) is used to estimate it as shown in Equation (2.42).

$$\hat{\beta}_0 = \bar{Y} - \hat{\beta}_1 \bar{X} \qquad (2.42)$$

Upon the estimation of β_0 and β_1, these coefficients can be used to make predictions about the dependent variable Y for new unseen data values of the independent variable X and modeled as shown in Equation (2.43).

$$\hat{Y} = \hat{\beta}_0 + \hat{\beta}_1 X \qquad (2.43)$$

ii. **Multivariate Linear Regression**
Multivariate linear regression involves more than one independent variable (predictor variable) to predict a single dependent variable. The general form of the multivariate linear regression model is shown in Equation (2.44).

$$Y = \beta_0 + \beta_1 X_1 + \beta_2 X_2 + \cdots + \beta_k X_k + \varepsilon \qquad (2.44)$$

where:

Y is the dependent variable.
$X_1, X_2, \ldots, X_k$ are independent variables (predictors).
$\beta_0, \beta_1, \beta_2, \ldots, \beta_k$ are the coefficients (intercepts and slopes) representing the relationship between each independent variable and the dependent variable.
ε is the error term, representing the difference between the observed and predicted values of Y ($\varepsilon = Y_i - \hat{Y}_i$).

In the real scenario the dataset will have multiple k features with n records; it can be modeled as shown in Equation (2.45).

$$\begin{aligned}
Y_1 &= \beta_0 + \beta_1 X_{11} + \beta_2 X_{12} + \cdots + \beta_k X_{1k} + \varepsilon_1 \\
Y_2 &= \beta_0 + \beta_1 X_{21} + \beta_2 X_{22} + \cdots + \beta_k X_{2k} + \varepsilon_2 \\
&\vdots \\
Y_n &= \beta_0 + \beta_1 X_{n1} + \beta_2 X_{n2} + \cdots + \beta_k X_{nk} + \varepsilon_n
\end{aligned} \qquad (2.45)$$

These n Equations from Equation (2.45) can be written as shown in Equation (2.46).

$$\begin{pmatrix} Y_1 \\ Y_2 \\ \vdots \\ Y_n \end{pmatrix} = \begin{pmatrix} 1 & X_{11} & X_{12} & \cdots & X_{1k} \\ 1 & X_{21} & X_{22} & \vdots & X_{2k} \\ \vdots & \vdots & \vdots & \ddots & \vdots \\ 1 & X_{n1} & X_{n2} & \cdots & X_{nk} \end{pmatrix} \begin{pmatrix} \beta_0 \\ \beta_1 \\ \vdots \\ \beta_k \end{pmatrix} + \begin{pmatrix} \varepsilon_1 \\ \varepsilon_2 \\ \vdots \\ \varepsilon_n \end{pmatrix} \qquad (2.46)$$

In general, for a multiple linear regression, the model with k independent features (variables) can be simply expressed as shown in Equation (2.47).

$$y = x\beta + \varepsilon \qquad (2.47)$$

where:

$$y = \begin{pmatrix} Y_1 \\ Y_2 \\ \vdots \\ Y_n \end{pmatrix}$$

$$x = \begin{pmatrix} 1 & X_{11} & X_{12} & \cdots & X_{1k} \\ 1 & X_{21} & X_{22} & \vdots & X_{2k} \\ \vdots & \vdots & \vdots & \ddots & \vdots \\ 1 & X_{n1} & X_{n2} & \cdots & X_{nk} \end{pmatrix}$$

$$\beta = \begin{pmatrix} \beta_0 \\ \beta_1 \\ \vdots \\ \beta_k \end{pmatrix} \text{ and }$$

$$\varepsilon = \begin{pmatrix} \varepsilon_1 \\ \varepsilon_2 \\ \vdots \\ \varepsilon_n \end{pmatrix}$$

To estimate (predict) the value of a dependent variable from Equation (2.47), it's necessary to estimate the parameters (regression coefficients) using the Ordinary Least Squares (OLS) method that minimizes the error term as shown in Equation (2.48).

$$L = \sum_{i=1}^{n} \varepsilon_i^2 = \varepsilon'\varepsilon = (y - x\beta)'(y - x\beta) \tag{2.48}$$

The resulting least squares estimate is shown in Equation (2.49).

$$\hat{\beta} = \left(x^T x\right)^{-1} x^T y \tag{2.49}$$

Since the estimation of dependent variable can be obtained by $\hat{y} = x\hat{\beta}$, then the equation of a multivariate linear regression model can be obtained as shown in Equation (2.50).

$$\hat{y} = x\left(x^T x\right)^{-1} x^T y \tag{2.50}$$

2.3 PROBABILITY THEORY

Probability theory is a branch of mathematics that studies random events and the likelihood of their occurrence. It provides a mathematical framework for quantifying uncertainty and predicting the probability of specific outcomes in events with multiple possible results. Understanding probability theory is crucial for machine learning practitioners, as it underpins many machine learning algorithms. The subsequent subsections offer a comprehensive overview of key probability concepts, helping readers build a solid foundation in probability theory and its applications in machine learning.

2.3.1 Sample spaces and events

Probability theory entails fundamental concepts like sample space, probability distributions, and random variables to calculate the likelihood of an event to occur. A sample space is the collection of all conceivable experiment results. It comprises all possible outcomes that could occur during an experiment. For example, when flipping a coin, the sample space has two possible outcomes: heads or tails. An event is a subset of the sample space that denotes specific outcomes or combinations of outcomes. Events can range from basic (like flipping a coin) to compound (like flipping a coin twice and getting heads both times). Understanding sample spaces and events is critical for comprehending probability and generating predictions in a variety of domains. This is where probability theory is heavily used to quantify uncertainty and make decisions. The following are the types of events:

   i. **Independent Events**
      These are the events that occur without being influenced by other factors. This implies that the outcome of one event does not affect the outcome of another.
  ii. **Dependent Events**
      These are events that are influenced by prior outcomes. This suggests that the occurrence of one event has a considerable impact on the probability of the succeeding event.
 iii. **Mutually Exclusive Events**
      These are events that are characterized by their inability to occur simultaneously. When one of these events takes place, the occurrence of the others is precluded.
  iv. **Equally Likely Events**
      These are events that share an identical probability of happening. This implies that, under similar conditions, each of the events has an equal chance of occurrence.
   v. **Exhaustive Events**
      These are events that encompass all possible outcomes within the sample space of an experiment. They essentially account for every conceivable result that could arise from the given set of circumstances.

2.3.2 Probability

Probability is defined as the ratio of the number of favorable outcomes to the total number of possible outcomes. Suppose S is the sample space, representing the set of all possible outcomes of an experiment and an event A is a subset of sample space S. Hence, the probability of an event A denoted as P(A) is defined as the ratio of the number of favorable outcomes for event A denoted as n(A) to the total number of possible outcomes in the sample space denoted as n(S). Mathematically, P(A) is computed as shown in Equation (2.51).

$$P(A) = \frac{n(A)}{n(S)} \tag{2.51}$$

2.3.3 Probability measures

A probability measure assigns numerical values to events within a sample space, reflecting the likelihood of occurrence of those events. It provides a formal framework for quantifying uncertainty and making predictions in various fields, including statistics and machine learning. A probability measure P on a sample space S satisfies the following properties:

i. **Non-negativity**
This property states that the probability of an event A must be a non-negative real number. In mathematical terms, the property is represented as $P(A) \geq 0$ for all events A.
ii. **Normalization**
This property states that the total probability assigned to the entire sample space is equal to 1. Mathematically, for a sample space S, the property is represented as $P(S) = 1$.
iii. **Additivity**
The additivity property of a probability measure applies to mutually exclusive events. The additivity property states that the probability of the union of two mutually exclusive events is equal to the sum of their individual probabilities. If $A_1, A_2, \ldots$ are disjoint events (i.e., $A_i \cap A_j = 0$ whenever $i \neq j$), then $P(A_1 \cup A_2 \cup \ldots) = \sum_i P(A_i)$.

2.3.4 Conditional probability

Conditional probability is a measure of the likelihood of an event occurring given that another event has already occurred with a certain probability. It is denoted by $P(A|B)$, where A represents the event of interest and B signifies the condition under consideration for evaluating the probability. Mathematically, conditional probability is computed as shown in Equation (2.52).

$$P(A|B) = \frac{P(A \cap B)}{P(B)} \tag{2.52}$$

where:

$P(A|B)$ is the conditional probability of event A given event B has occurred.
$P(A \cap B)$ is the joint probability of events A and B occurring together. If the two events (A and B) are independent (i.e., mutually exclusive events), then $P(A \cap B) = P(A)P(B)$.

Thus, the conditional probability becomes $P(A \mid B) = P(A)$. This is equivalent to stating that the observation of B has no impact on the probability of A.
$P(B)$ is the probability of event B occurring.

2.3.5 Bayes' theorem

Bayes' Theorem is an important concept in probability theory that offers a method for updating beliefs about the probability of an event occurring based on new evidence. It is a cornerstone in various fields such as statistics, machine learning, and AI. Mathematically, the theorem relates the conditional probability of an event A given event B (i.e., the posterior probability) to the conditional probability of event B given event A (i.e., the likelihood), along with the prior probabilities of events A and B occurring independently. The formula of Bayes' Theorem is given in Equation (2.53).

$$P(A \mid B) = \frac{P(B \mid A) \cdot P(A)}{P(B)} \tag{2.53}$$

where:

$P(A \mid B)$ is the conditional probability of event A given event B has occurred.
$P(B \mid A)$ is the conditional probability of event B given event A has occurred.
$P(A)$ is the probability of event A occurring.
$P(B)$ is the probability of event B occurring.

2.3.6 Random variables

A random variable is a mathematical function that assigns a numerical value to each possible outcome of a random experiment. In simpler terms, a random variable is a variable whose value is determined by the outcome of a random process. There are mainly two types of random variables: discrete and continuous, as described in the following.

   i. **Discrete Random Variables**
      These are variables that take on a countable number of distinct values. The possible values of a discrete random variable can be listed, and there are gaps between them. Examples of random variables include the number of heads in a series of coin flips or the count of emails received in a day.
   ii. **Continuous Random Variables**
      These are variables that can take any value within a given range. The possible values form a continuous interval, and there are no gaps between them. Examples include the height of individuals in a population, the time it takes for a reaction to occur, or the temperature at a specific location.
      Random variables are denoted by X and their possible values are often denoted by lowercase letters, e.g., x. The probability distribution of a random variable describes the likelihood of each possible value occurring.

2.3.7 Expectation

Expectation or mean represents the average value that one would expect the random variable X to take over a large number of repetitions of an experiment. It is denoted by $E(x)$ or μ and is a measure of central tendency.

For a discrete random variable X, with probability mass function $P(X = x_i)$ and corresponding values x_i, the expectation is denoted in Equation (2.54).

$$E(X) = \sum_i x_i \cdot P(X = x_i) \tag{2.54}$$

For a continuous random variable X, with probability density function $f(x)$, the expectation is denoted in Equation (2.55).

$$E(X) = \int_{-\infty}^{\infty} x \cdot f(x) dx \tag{2.55}$$

2.3.8 Variance

The variance of a random variable X is a measure of the spread or dispersion of its values around the mean or expected value. It quantifies the degree to which individual observations deviate from the average. Variance is denoted as Var(x) or σ^2.

For a discrete random variable X, with probability mass function $P(X = x_i)$ and corresponding values x_i and expected value μ, the variance is calculated as indicated in Equation (2.56).

$$\text{Var}(X) \text{ or } \sigma^2 = \sum_i (x_i - \mu)^2 \cdot P(X = x_i) \tag{2.56}$$

Whereas, for a continuous random variable X, with probability density function $f(x)$ and expected value μ, the variance is computed as shown in Equation (2.57).

$$\text{Var}(X) \text{ or } \sigma^2 = \int_{-\infty}^{\infty} (x - \mu)^2 \cdot f(x) dx \tag{2.57}$$

2.3.9 Standard deviation

The standard deviation is a statistical measure that quantifies the amount of variation or dispersion within a set of values. Standard deviation is the square root of variance. Additionally, it shows how individual data points deviate from the dataset's mean. Hence, a low standard deviation shows that the data points are close to the mean, whereas a high standard deviation indicates more variability. Standard deviation is denoted by σ.

For a discrete random variable X, with probability mass function $P(X = x_i)$ and corresponding values x_i and expected value μ, the standard deviation is given in Equation (2.58).

$$\sigma = \sqrt{\sum_i (x_i - \mu)^2 \cdot P(X = x_i)} \quad \text{or} \quad \sigma = \sqrt{\frac{\sum_{i=1}^n (x_i - \mu)^2}{n}} \tag{2.58}$$

Equation (2.59) denotes the variance for a continuous random variable X, with probability density function $f(x)$ and expected value μ.

$$\sigma = \sqrt{\int_{-\infty}^{\infty} (x - \mu)^2 \cdot f(x) dx} \tag{2.59}$$

A probability distribution is a mathematical function that describes the probability of various outcomes in a random experiment. It offers a way for assigning probabilities to the numerous outcomes that a random variable can have. Understanding probability distributions is essential for probability theory, statistics, machine learning, and data science. Furthermore, when describing probability measures linked with random variables, alternative functions such as cumulative distribution functions (CDFs), probability density functions (PDFs), and probability mass functions (PMFs) are frequently defined. These functions provide a straightforward approach to calculating the probability measure that will lead an experiment.

2.3.9.1 Cumulative distribution function

The cumulative distribution function (CDF) depicts the probability distribution of a random variable. It also provides the likelihood that the variable will have a value less than or equal to a given value x. Consider the random variable X, which represents the adult male height in a population as measured in feet. The CDF of X, denoted by $f(x)$, indicates the probability that an adult male is less than or equal to a given height, such as 68 feet. Specifically, $f(68)$ is the probability that an adult male is shorter than or equal to 68 feet. For any random variable X, the CDF $f(x)$ must meet the following conditions.

i. **Non-decreasing**: This feature means that when x increases, the cumulative probability does not decrease. If $f(x_1) \leq f(x_2)$, then $x_1 \leq x_2$. Where $f(x_1)$ and $f(x_2)$ denote the CDF values at two points x_1 and x_2, respectively.
ii. **Right-Continuous**: The probability of reaching any particular value from the right is the same as approaching it from the left. It is denoted as $F(x) = \lim_{h \to 0^+} F(x+h)$.
iii. **Limits at Infinity**: The cumulative probability approaches 0 for values that are extremely small and approaches 1 for values that are extremely large. It is denoted as $\lim_{x \to -\infty} F(x) = 0$ and $\lim_{x \to +\infty} F(x) = 1$.

2.3.9.2 Probability mass function

The probability mass function (PMF) represents the probability distribution of a discrete random variable by assigning probabilities to all possible outcomes. Thus, the PMF is denoted as $P(X = x)$, and it reflects the probability that the random variable X will take the value x. For a discrete random variable X, the PMF $P(X = x)$ satisfies non-negativity (i.e., $P(X) \geq 0$) and summation to 1 (i.e., $\sum_{\text{all } x} P(X = x) = 1$) property. The PMF is significant in machine learning because it provides a formal mechanism for describing the probability distribution of discrete random variables. This allows machine learning systems to represent and reason about uncertainty in discrete domains.

2.3.9.3 Probability density function

A Probability Density Function (PDF) defines the probability distribution of a continuous random variable by assigning probabilities to groups of values rather than individual values. Furthermore, the PDF, indicated as $f(x)$, represents the likelihood that the random variable X falls inside a specific range around x. Consider X, a continuous random variable representing an adult male's height in a population, measured in feet. Thus, the PDF $f(x)$ would represent the likelihood that an adult male's height falls within a specified range of x feet. A PDF $f(x)$ must satisfy two conditions: non-negativity ($f(x) \geq 0$) and area under the curve ($\int_{-\infty}^{\infty} f(x)dx = 1$). The PDF is significant in machine learning because it allows for formal modeling and analysis of continuous random variables. This allows algorithms to understand uncertainty and make accurate predictions in continuous domains.

2.3.9.4 Discrete distributions

Discrete probability distributions show the probabilities associated with discrete random variables, which have separate and independent values. Discrete distributions include the Bernoulli, Binomial, and Poisson distributions. These sorts are ideal for modeling events with countable and accurate outcomes.

2.3.9.5 Bernoulli distribution

The Bernoulli distribution represents a random experiment with only two possible outcomes (1 for success and 0 for failure), making it ideal for representing binary data. It is useful in machine learning, particularly for classification tasks. Furthermore, Equation (2.60) defines the probability mass function, $P(X = k)$, of a Bernoulli random variable X. In addition, Equations (2.61) and (2.62) calculate the mean, $E(X)$, and variance, $Var(X)$, of the Bernoulli Distribution, respectively.

$$P(X = k) = p^k \cdot (1-p)^{1-k} \qquad (2.60)$$

where:

k takes values 0 or 1, and

p represents the probability of success. The distribution is characterized by a single parameter p, which is between 0 and 1.

$$E(X) = p \tag{2.61}$$

$$\text{Var}(X) = p \cdot (1-p) \tag{2.62}$$

2.3.9.6 Binomial distribution

The number of successes in a fixed number of independently and identically distributed Bernoulli trials is represented by a binomial distribution. Each trial either succeeds with probability p or fails with probability $(1 - p)$. Equation (2.63) describes the probability mass function for a binomial random variable X. Equations (2.64) and (2.65) also provide the mean, $E(X)$, and variance, $\text{Var}(X)$, of a Binomial Distribution.

$$P(X = k) = \binom{k}{n} \cdot p^k \cdot (1-p)^{n-k} \tag{2.63}$$

where:

n is the number of trials.
k is the number of successes.
p is the probability of success in a single trial.
$\binom{k}{n}$ is the binomial coefficient, representing the number of ways to choose k success from n trials.

$$E(X) = np \tag{2.64}$$

$$\text{Var}(X) = np(1-p) \tag{2.65}$$

2.3.9.7 Poisson distribution

The Poisson Distribution represents the number of events that occur within a specific time. This distribution is crucial in machine learning for modeling unusual event occurrences within a set period, which aids in tasks such as website traffic prediction and data anomaly detection. Equation (2.66) calculates the PMF for a Poisson random variable X. In addition, the mean $E(X)$ and variance $\text{Var}(X)$ of a Poisson distribution are equal to the average rate parameter λ, as defined in Equation (2.67).

$$P(X = k) = \frac{\lambda^k e^{-\lambda}}{k!} \tag{2.66}$$

where:

k is the number of events.
λ is the average rate at which events occur.

e is the base of the natural logarithm (i.e., $e \approx 2.71828$).

$$E(X) = \text{Var}(X) = \lambda \tag{2.67}$$

2.3.9.8 Uniform distribution

The Uniform Distribution is distinguished by a PMF that is constant throughout a specified range. The distribution is uniform since all outcomes within the range have an equal chance of occurring. Furthermore, in machine learning, the distribution is critical for producing random samples with similar probability across a certain range and giving a baseline comparison. As a result, it is critical in producing synthetic datasets for model training and testing, as well as in assuring random selection process integrity. It can also generate random starting settings for algorithms. Equation (2.68) computes the PMF for a uniform random variable X on the interval $[a, b]$. For a uniform distribution, the mean $E(X)$ and variance $\text{Var}(X)$ are also calculated using Equations (2.69) and (2.70).

$$P(X = x_i) = \frac{1}{b-a+1} \tag{2.68}$$

where:

a is the minimum value in the range.
b is the maximum value in the range.
x_i is a specific value in the range.

$$E(X) = \frac{a+b}{2} \tag{2.69}$$

$$\text{Var}(X) = \frac{(b-a+1)^2 - 1}{12} \tag{2.70}$$

2.3.9.9 Continuous distributions

Continuous probability distributions describe the probabilities associated with continuous random variables. Unlike discrete distributions, where the random variable can only assume distinct values, continuous distributions deal with variables that can take on an uncountable infinite number of values within a given interval. These distributions are vital for machine learning in modeling real-world phenomena with continuous variables and facilitate tasks such as regression, density estimation, and generative modeling. Types of continuous probability distribution include Normal Distribution, Uniform Distribution, Exponential Distribution, Log-Normal Distribution, Gamma Distribution, and Beta Distribution.

2.3.9.10 Normal distribution (Gaussian distribution)

The Normal Distribution, commonly referred to as the Gaussian distribution, is a foundational probability distribution well-known for its symmetry about the mean. This inherent symmetry implies that data points close to the mean are more prevalent than those farther away, creating the distinctive bell-shaped curve appearance when visualized graphically. This distribution is widely used for data analysis, anomaly detection, and generating synthetic data in machine learning. The probability density function for a Gaussian random variable X with a mean $E(X)$ and standard deviation (σ) is given in Equation (2.71). Moreover, the mean $E(X)$ and variance $\text{Var}(X)$ for a Normal Distribution are calculated as shown in Equations (2.72) and (2.73), respectively.

$$f(x \mid \mu, \sigma) = \frac{1}{\sqrt{2\pi\sigma^2}} e^{-\frac{(x-\mu)^2}{2\sigma^2}} \qquad (2.71)$$

where:

x is the random variable.
μ is the mean, determining the center of the distribution.
σ is the standard deviation, influencing the spread or dispersion of the distribution.
π is the mathematical constant (i.e., $\pi \approx 3.14159$).
e is the base of the natural logarithm ($e \approx 2.71828$).

$$E(X) = \mu \qquad (2.72)$$

$$\text{Var}(X) = \sigma^2 \qquad (2.73)$$

2.3.9.11 Uniform distribution

The Uniform Distribution is a probability distribution characterized by a constant PDF over a specified range. In simpler terms, every outcome within the range has an equal chance of occurring, making the distribution uniform. This distribution is important in machine learning for fair random selection processes and a crucial tool in generating random samples that can be used in algorithm training and testing. The PDF for a uniform random variable X over the interval $[a,b]$ is given in Equation (2.74). Moreover, the mean $E(X)$ and variance $\text{Var}(X)$ for a uniform distribution are calculated as shown in Equations (2.75) and (2.76), respectively.

$$f(x \mid a,b) = \frac{1}{(b-a)} \qquad (2.74)$$

where:

a is the lower bound of the interval.
b is the upper bound of the interval.

$$E(X) = \frac{a+b}{2} \qquad (2.75)$$

$$\text{Var}(X) = \frac{(b-a)^2}{12} \qquad (2.76)$$

2.4 CALCULUS

Calculus is essential in machine learning, particularly for optimizing algorithms and understanding function behavior. Consequently, differentiation and integration are two fundamental concepts that are often employed in machine learning.

2.4.1 Differentiation

Differentiation is used to determine the rate at which a function changes. In machine learning, it is commonly used to optimize models by modifying parameters to reduce or maximize a specific objective function. The derivative of a function $f(x)$ with respect to a variable x is represented by $f'(x)$, which represents the rate of change of $f(x)$ at a particular position. The derivative is defined as the limit of the difference quotient as the interval approaches 0, as shown in Equation (2.77).

$$f'(x) = \lim_{h \to 0} \frac{f(x+h) - f(x)}{h} \tag{2.77}$$

Let us consider a simple function $f(x) = x^2$. Its derivative $f'(x)$ can be computed using the power rule of differentiation:

$$f'(x) = 2x, \text{ which is the derivative of } f(x) = x^2.$$

2.4.2 Integration

Integration is the reverse process of differentiation; hence, it is used to calculate the area under a function's curve. In machine learning, integration is used in a range of situations, including predicting probabilities in statistical models. Equation (2.78) represents the integral of a function $f(x)$ with relation to the variable x.

$$\int f(x) dx \tag{2.78}$$

Let us consider the function $g(x) = 2x$. The area under the curve can be calculated by integrating $g(x)$ with respect to x.

$$\int f(x) dx = x^2 + C$$

where:

C is the integration constant.

To get the area under the curve of $g(x) = 2x$ from $x = 0$ to $x = 3$, we use the definite integral:

$$\int_0^3 2x\,dx = \left[x^2\right]_0^3 = 3^2 - 0^2 = 9$$

So, the area under the curve of $g(x) = 2x$ from $x = 0$ to $x = 3$ is 9.

2.4.3 Gradient

Gradients are a basic concept in calculus that play a significant role in model optimization by providing information about the rate of change of functions. To improve a model's performance, a cost function that measures the difference between the model's predictions and the desired outcome is usually minimized. In addition, several machine learning optimization approaches, such as gradient descent and its derivatives, iteratively update model parameters using the gradient. As a result, in order to achieve optimal performance, a model's internal parameters must be modified so that the cost gradually decreases. Here is when the concept of gradient comes into play. In mathematics, gradients are the vectors of partial derivatives of a multivariable function with respect to its input variables. Geometrically, the gradient indicates the direction of the steepest ascent of a function's surface at a particular point. Equation (2.79) defines the gradient of a function $f(x)$.

$$\nabla f = \left(\frac{\partial f}{\partial x_1}, \frac{\partial f}{\partial x_2}, \ldots, \frac{\partial f}{\partial x_n}\right) \qquad (2.79)$$

where:

$\frac{\partial f}{\partial x_i}$ represents the partial derivative of f with regard to the i-th input variable, x_i.

2.4.4 Linear function

A linear function is a mathematical relationship between two variables that can be represented visually by a straight line, with the dependent variable moving at a constant rate relative to the independent variable. Let us consider the simple linear function $f(x) = mx + c$, where m is the slope and c is the intercept. As a result, the gradient of f with respect to x is constant, equal to the slope m across the domain. Hence, the gradient of f is $\nabla f = m$. For example, given a linear function $f(x) = 2x + 3$, the gradient of f with respect to x is constant and equal to $m = 2$ across the domain. Therefore, ∇f equals 2 for all x.

2.4.5 Quadratic function

A quadratic function is a mathematical relationship between two variables that can be graphically depicted as a curve in which the dependent variable increases or decreases in proportion to the independent variable squared. Let us consider a quadratic function

$f(x) = ax^2 + bx + c$, where a, b, and c are constants. When partial derivatives are computed, the gradient ∇f varies with x and is influenced by coefficients a and b. The gradient vector provides insight on the slope of the quadratic curve at various points within its domain. Consider the quadratic function $f(x) = x^2 + 2x + 1$. Its gradient with respect to x is $\nabla f = 2x + 2$. At $x = 0$, the gradient is $\nabla f = 2$, showing a positive slope. As x increases, so does the gradient, which reflects the quadratic curve's steeper slope.

2.4.6 Sigmoid function

A sigmoid function is a mathematical function with a distinctive S-shaped curve. It is commonly used to model nonlinear interactions and map inputs to a range of 0 to 1. In machine learning, the sigmoid function $\sigma(x) = 1/(1 + e^{-x})$ is frequently employed as an activation function. The gradient of the sigmoid function with respect to x has a distinct S-shape, gradually shifting from big positive values to small positive values as x changes. Let us get the gradient of the sigmoid function $\sigma(x) = 1/(1 + e^{-x})$ with respect to x. Calculating the derivative of $\sigma(x)$ yields $\sigma'(x) = \sigma(x)(1 - \sigma(x))$. This gradient goes smoothly from big positive values (i.e., for large positive x) to small positive values (i.e., for large negative x), representing the sigmoid function's characteristic of translating input values to the range (0, 1).

2.5 GEOMETRY AND TRIGONOMETRY

Geometry and trigonometry are fundamental mathematical principles with numerous applications in machine learning, ranging from data representation to model development. Understanding their applications is critical for developing efficient algorithms and evaluating their outcomes. This section looks into the fundamental ideas of geometry and trigonometry, emphasizing their application in the machine learning domain.

2.5.1 Geometry in data representation

Geometry provides a foundation for describing data in machine learning, especially in high-dimensional domains. Distance metrics, inner products, and norms are important concepts for assessing the similarities and differences between data points. For example, the Euclidean distance metric calculates the straight-line distance between points in a geometric space, making it easier to cluster, classify, and discover anomalies. Consider a dataset with two-dimensional points reflecting the positions of houses in a neighborhood. Each data point (x, y) represents the coordinates of a dwelling on a map. Equation (2.80) is used to calculate the Euclidean distance between pairs of data points to determine how similar dwellings are based on their locations.

$$\text{Distance} = \sqrt{(x_2 - x_1)^2 + (y_2 - y_1)^2} \tag{2.80}$$

Let us consider two houses with coordinates (2, 3) and (5, 7). The Euclidean distance between them can be computed as follows:

$$\text{Distance} = \sqrt{(5-2)^2 + (7-3)^2} = \sqrt{3^2 + 4^2} = \sqrt{9+16} = \sqrt{25} = 5$$

Therefore, the Euclidean distance between the dwellings is 5 units. This distance measure allows us to quantify the spatial links between dwellings and perform tasks like clustering or identifying nearest neighbors for recommendation systems.

2.5.2 Trigonometric geometry in model optimization

Trigonometric functions, particularly hyperbolic functions such as hyperbolic sine (sinh), cosine (cosh), and tangent (tanh), are important in model optimization and activation functions in machine learning. To introduce non-linearity and assist gradient-based optimization, neural networks frequently use hyperbolic tangent (tanh) and rectified linear units (ReLU) activation functions. To incorporate non-linear transformations and normalize activations in deep neural networks, the hyperbolic tangent (tanh) activation function is used to neuron outputs. This enables more effective gradient propagation and convergence during back-propagation, resulting in better training stability and model performance. Assume there is a neural network consisting of one input neuron, one hidden neuron, and one output neuron. The hidden neuron will utilize the hyperbolic tangent activation function, whereas the output neuron will use the rectified linear unit (ReLU) activation function.

The hyperbolic tangent function, denoted as $\tanh(x)$, is defined in Equation (2.81).

$$\tanh(x) = \frac{\sinh(x)}{\cosh(x)} = \frac{e^x - e^{-x}}{e^x + e^{-x}} \qquad (2.81)$$

where:

$$\sinh(x) = \frac{e^x - e^{-x}}{2} \text{ and}$$

$$\cosh(x) = \frac{e^x + e^{-x}}{2}$$

Given an input x, the hidden neuron computes its output h using the hyperbolic tangent activation function, as illustrated in Equation (2.82).

$$h = \tanh(wx + b) \qquad (2.82)$$

where:

w represents the weight connecting the input to the hidden neuron, and b is the bias term.

Equation (2.83) shows how the output neuron computes its output y using the rectified linear unit (ReLU) activation function.

$$y = \max(0, \text{wh} + c) \tag{2.83}$$

where:

wh represents the weighted sum of the hidden output of the neuron, and c is the output bias term of a neuron.

Assume that the weight linking the input to the hidden neuron is $w = 0.5$, the bias term for the hidden neuron is $b = 1$, the weight connecting the hidden neuron to the output neuron is $w' = -1$, and the bias term for the output neuron is $c = 0.5$. Given an input $x = 2$, we can compute the output of the hidden neuron using Equation (2.82) as follows:

$$h = \tanh(0.5 \times 2 + 1) = \tanh(2 + 1) = \tanh(3)$$

Using the hyperbolic tangent function, we determine that h is around 0.995. The output neuron's output is then calculated using Equation (2.83), as follows:

$$y = \max(0, -1 \times 0.995 + 0.5) = \max(0, -0.995 + 0.5) = \max(0, -0.495) = 0$$

Therefore, the output of the output neuron is 0.

2.6 INFORMATION THEORY

Information theory is an area of mathematics created in 1948 by Claude Shannon that provides a framework for quantifying and studying information, uncertainty, and communication systems. In the context of machine learning, information theory provides important insights into data representation, model evaluation, and optimization techniques. This subsection looks into the fundamental concepts of information theory and its applications in machine learning, with instructive examples.

2.6.1 Entropy and information content

Entropy is a fundamental concept in information theory that describes the average uncertainty or disorder in a probability distribution. It estimates the quantity of information needed to describe the results of a random variable. In machine learning, entropy is an important statistic for assessing uncertainty in data distributions and model predictions. For example, in decision tree algorithms, entropy is used to assess the purity of splits and drive feature selection. The entropy $H(X)$ of a discrete random variable X with probability distribution $P(X)$ is computed using Equation (2.84).

$$H(\mathcal{X}) = -\sum_{x \in \mathcal{X}} P(x) \log_2 P(x) \qquad (2.84)$$

where:

x is the set of all possible values for X.

Assume we have a random variable X that represents the result of flipping a fair coin. There are two possible outcomes: heads (H) and tails (T), each with a chance of 0.5. This distribution's entropy is computed using Equation (2.84):

$$H(\mathcal{X}) = -\left(\frac{1}{2}\log_2 \frac{1}{2} + \frac{1}{2}\log_2 \frac{1}{2}\right) = -\left(\frac{1}{2} - \frac{1}{2}\right) = 1 \text{ bit}$$

This indicates that there is 1 bit of uncertainty associated with each coin-flip outcome.

For example, in classification, consider a binary classification problem with two classes, where each class occurs with an equal probability ($p = 0.5$). The entropy of this distribution is calculated as shown in Equation (2.84).

$$\begin{aligned}\text{Entropy} &= -p_1 \log_2 p_1 - p_2 \log_2 p_2 \\ &= -0.5\log_2(0.5) - 0.5\log_2(0.5) = -0.5 \times (-1) - 0.5 \times (-1) = 1 \text{ bit}\end{aligned}$$

This indicates that there is 1 bit of uncertainty associated with each outcome, reflecting the equal probability of the two classes.

2.6.2 Mutual information and feature selection

Mutual information measures the amount of information shared between two random variables. In machine learning, mutual information is utilized for feature selection, where it quantifies the relevance of each feature to the target variable. Features with high mutual information are considered informative and are retained, while irrelevant features are discarded. Mutual Information $I(X;Y)$ between two random variables X and Y with joint probability distribution $P(X,Y)$ is calculated as shown in Equation (2.85).

$$I(X;Y) = \sum_{x \in X}\sum_{y \in Y} P(x,y) \log_2\left(\frac{P(x,y)}{P(x)P(y)}\right) \qquad (2.85)$$

where:

x and y are the sets of possible values of X and Y, respectively.

Consider a dataset with two variables X and Y, where X represents the presence (i.e., 1) or absence (i.e., 0) of a particular gene mutation and Y represents the occurrence

TABLE 2.1 The values of features X and Y

X	Y	COUNT
0	0	500
0	1	200
1	0	100
1	1	600

(1) or absence (0) of a disease as shown in Table 2.1. The aim is to measure the mutual information between X and Y to determine the relevance of the gene mutation to the disease.

Using the formula for mutual information in Equation (2.85), the following calculation can be performed.

$$I(X;Y) = P(0,0)\log_2\left(\frac{P(0,0)}{P(0)P(0)}\right) + P(0,1)\log_2\left(\frac{P(0,1)}{P(0)P(1)}\right)$$
$$+ P(1,0)\log_2\left(\frac{P(1,0)}{P(1)P(0)}\right) + P(1,1)\log_2\left(\frac{P(1,1)}{P(1)P(1)}\right)$$
$$= \frac{500}{1400}\log_2\left(\frac{500}{1400\times 600}\right) + \frac{200}{1400}\log_2\left(\frac{200}{1400\times 800}\right)$$
$$+ \frac{100}{1400}\log_2\left(\frac{100}{1400\times 600}\right) + \frac{600}{1400}\log_2\left(\frac{600}{1400\times 800}\right) \approx 0.041\,\text{bits}$$

This indicates the amount of information gained about the disease (Y) by observing the gene mutation (X), with higher values indicating a stronger association.

2.6.3 Cross-entropy and model evaluation

Cross-entropy is a measure of dissimilarity between two probability distributions. In machine learning, it is commonly used as a loss function for training classification models, particularly in neural networks. Minimizing cross-entropy corresponds to maximizing the likelihood of predicting the correct class label. For example, in binary classification, the cross-entropy loss function is defined in Equation (2.86).

$$\text{Cross} - \text{Entropy} = -\frac{1}{N}\sum_{i=1}^{N}\left[y_i \log p_i + (1 - y_i)\log(1 - p_i)\right] \quad (2.86)$$

where:

y_i is the true class label (0 or 1),
p_i is the predicted probability of the positive class, and
N is the number of samples.

2.7 CLUSTERING

As explained earlier in Chapter 1, clustering identifies patterns in unlabeled data by grouping similar data points into clusters or segments. Although there are several categories of clustering algorithms, for mathematical illustrative purposes, the *K*-Means clustering algorithm based on partitioning clustering is demonstrated in the subsequent subsection.

2.7.1 *K*-Means clustering algorithm

The *K*-Means algorithm is an iterative clustering technique used to partition a dataset into *K* distinct, non-overlapping clusters based on a specific distance metric (e.g., Euclidean distance). It works by iteratively assigning data points to the nearest cluster centroid and then updating the centroids based on the mean of the data points assigned to each cluster. This process continues until convergence, where the centroids no longer change significantly or a specified number of iterations are reached. The *K*-Means clustering algorithm can be performed through the following steps.

i. **Initialization**
   In the initial step, the parameter *K* is determined, representing the desired number of clusters. Subsequently, centroids are randomly initialized for each *K* cluster to start the clustering process.
ii. **Calculation of Distances**
   In this step, the distance matrix between the centroids and the data patterns should be created to identify the nearest distance of the data points to the centroids. Since there are *K* clusters/centroids and *n* samples, the algorithm shall compute *n*K* geometric distances. There are several geometric distances that can be used to compute the distance of the data points to the centroids, including the Euclidean distance, Manhattan distance, and Chebyshev distance. These geometric distances, together with their respective formulas are discussed in the following subsections.
   a. **Euclidean Distance**
      The Euclidean distance *d* is a straight-line distance between two points in a Euclidean space. It is computed using Equation (2.87).

$$d = \sqrt{\sum_{i=1}^{n}(x_i - y_i)^2} \qquad (2.87)$$

where:

n is the number of dimensions in Euclidean space
x_i and y_i are points in the Euclidean space

For example, given two points (2, 3) and (5, 7) in a two-dimensional space, as shown in Figure 2.8, the Euclidean distanc is computed as follows:

$$d = \sqrt{(2-5)^2 + (3-7)^2} = \sqrt{9+16} = \sqrt{25} = 5.$$

b. **Manhattan Distance**

This is the distance between two points in a grid-based system like a chessboard. It is calculated by adding the absolute differences of their coordinates using the formula in Equation (2.88).

$$d = \sum_{i=1}^{n} |x_i - y_i| \tag{2.88}$$

where:

x_i and y_i are points in the Euclidean space

Consider the same two points (2, 3) and (5, 7) in a two-dimensional space as shown in Figure 2.9 the Manhattan distance is computed as follows:

$$d = |2-5| + |3-7| = 3 + 4 = 7.$$

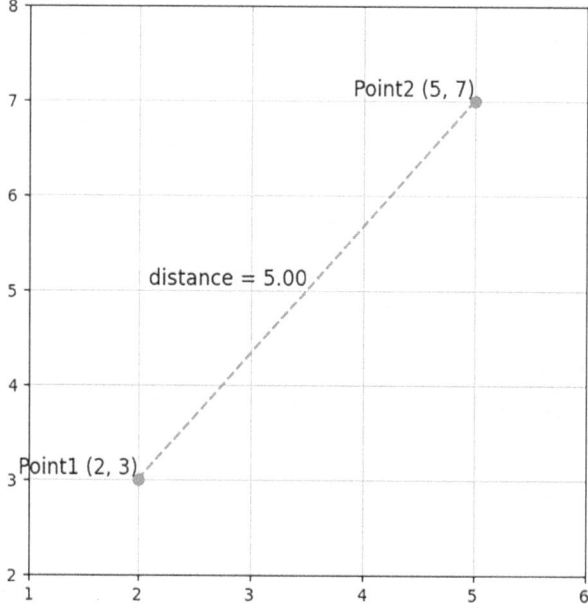

FIGURE 2.8 Euclidean distance visualization.

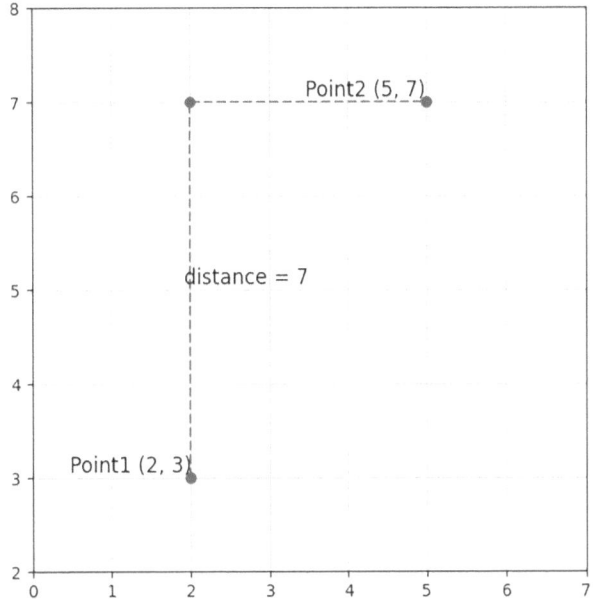

FIGURE 2.9 Manhattan distance visualization.

c. **Chebyshev Distance**
   The Chebyshev distance is the maximum absolute difference between two points across all dimensions. It is calculated using the formula in Equation (2.89).

$$d = \max_i \left(|x_i - y_i| \right) \tag{2.89}$$

where:

x_i and y_i are points in the Euclidean space.
For the same two points (2, 3) and (5, 7) in a two-dimensional space, as shown in Figure 2.10, the Chebyshev distance is computed as follows:

$$d = \max \left(|2-5|, |3-7| \right) = \max(3,4) = 4.$$

iii. **Assigning Each Sample in the Cluster**
   After calculating the distance from each sample to every cluster, the sample is assigned to the closest centroid (i.e., minimal distance). If a sample distance to the current centroid is much higher than one of the other centroid, then the sample should be shifted to the new centroid with minimum distance. However, when there is no movement of samples to another cluster anymore, the algorithm should end. Suppose the assignment of data points to centroids is determined

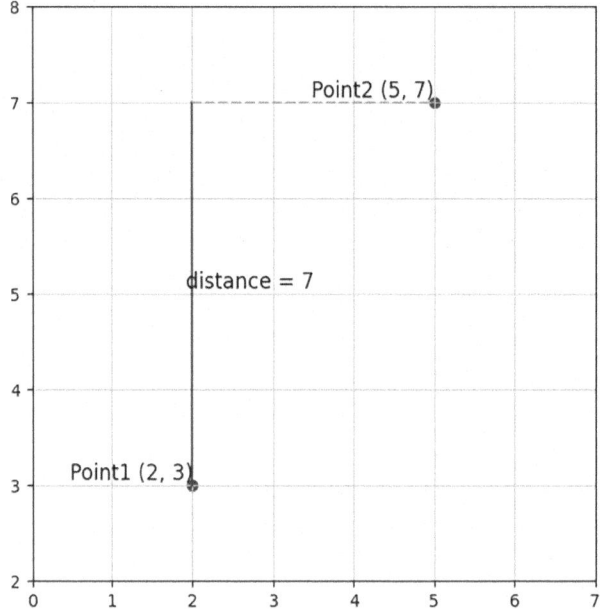

FIGURE 2.10 Chebyshev distance visualization.

using the Euclidean distance, then this can be performed using the formula in Equation (2.90).

$$d(x,c) = \sqrt{\sum_{i=1}^{n}(x_i - c_i)^2} \qquad (2.90)$$

where:

x is the data point
c is the centroid
n is the number of dimensions

iv. **Updation**

The updation step involves recalculating the centroids for each cluster by taking the mean of all data points assigned to the cluster. This will result in a shift in the positions of the centroids. The new centroid $c(c_x, c_y)$ is obtained as shown in Equations (2.91) and (2.92) for the x and y coordinates of data points, respectively.

$$c_x = \frac{1}{l}\sum_{i=1}^{l} x_i \qquad (2.91)$$

$$c_y = \frac{1}{l}\sum_{i=1}^{l} y_i \qquad (2.92)$$

v. **Repeating Steps ii to iv**

This step involves repeating steps *ii* to *iv* until the algorithm convergence (i.e., when the centroids no longer change) or a specified number of iterations is reached. Consider a dataset with eight samples and two attributes, as shown in Table 2.2. The task is to assign each data point to one of three clusters ($C1$, $C2$, and $C3$) using the *K*-Means algorithm, with the Manhattan distance serving as the distance measure. As described earlier in this section, the following steps are applied to assign each data point to the respective cluster as follows.

i. **Initialization Iteration 1**:

Since there are three clusters (i.e., $C1$, $C2$, $C3$), then the centroids are randomly initialized as follows:

$C1 : (A1 = 4.00, A2 = 6.33)$

$C2 : (A1 = 6.00, A2 = 5.67)$

$C3 : (A1 = 2.50, A2 = 5.50)$

ii. **Calculation of Distances and Assigning Each Sample to a Cluster**

Using the Manhattan distance measure, the distance from each data point to each centroid is calculated as follows.

Distances for data point (2, 10) to each centroid:

Distance to $C1 = |2 - 4| + |10 - 6.33| = 5.67$

Distance to $C2 = |2 - 6| + |10 - 5.67| = 8.33$

Distance to $C3 = |2 - 2.5| + |10 - 5.5| = 5$

Data point (2, 10) is clustered in $C3$ since it has the smallest distance of 5 from $C3$ compared to other clusters (5.67 and 8.33).

TABLE 2.2 Sample dataset

NO.	A1	A2
1	2	10
2	2	5
3	8	4
4	5	8
5	7	5
6	6	4
7	1	2
8	4	9

Distances for data point (2, 5) to each centroid:

Distance to $C1 = |2-4| + |5-6.33| = 3.33$

Distance to $C2 = |2-6| + |5-5.67| = 4.67$

Distance to $C3 = |2-2.5| + |5-5.5| = 1$

Data point (2, 5) is clustered in $C3$ since it has the smallest distance of 1 from $C3$ compared to other clusters (3.33 and 4.67).

Distances for data point (4, 8) to each centroid:

Distance to $C1 = |8-4| + |4-6.33| = 6.33$

Distance to $C2 = |8-6| + |4-5.67| = 3.67$

Distance to $C3 = |8-2.5| + |4-5.5| = 7$

Data point (4, 8) is clustered in $C2$ since it has the smallest distance of 3.67 from $C2$ compared to other clusters (6.33 and 7).

The assignment of data points to their respective clusters is shown in Table 2.3 with each color indicating the data points that belong to the same cluster.

iii. **Updating Centroids for Iteration 2**:
After assigning all samples to clusters, the centroids are recomputed by finding the mean of all data points in each cluster. The updated centroids will be used in the next iteration and are calculated as follows:

Centroid for Cluster 1, $C1$:

TABLE 2.3 Data points assigned to the clusters for the first iteration

				C1	C2	C3
			A1	4.00	6.00	2.50
DATA POINTS			A2	6.33	5.67	5.50
NO.	A1	A2		MANHATTAN DISTANCES		
1	2	10		5.67	8.33	5.00
2	2	5		3.33	4.67	1.00
3	8	4		6.33	**3.67**	7.00
4	5	8		2.67	3.33	5.00
5	7	5		4.33	1.67	5.00
6	6	4		4.33	1.67	5.00
7	1	2		7.33	8.67	5.00
8	4	9		2.67	5.33	5.00

- The data samples are (5, 8) and (4, 9).
- The mean is $\left(\frac{5+4}{2}, \frac{8+9}{2}\right) = (4.5, 8.5)$. Thus, the new centroid is $(4.5, 8.5)$.

Centroid for Cluster 2, C2:

- The data samples are (4, 8), (5, 7), and (4, 6).
- The mean is $\left(\frac{8+7+6}{3}, \frac{4+5+4}{2}\right) = (7, 4.33)$. Thus, the new centroid is $(7, 4.33)$.

Centroid for Cluster 3, C3:

- The data samples are (2, 10), (2, 5), and (1, 2).
- The mean is $\left(\frac{2+2+1}{3}, \frac{10+5+2}{2}\right) = (1.67, 5.67)$. Thus, the new centroid is $(1.67, 5.67)$.

iv. **Repeating Steps ii to iv (i2 to i3)**

The process of creating data point distances from each centroid, assigning data points to clusters, and updating centroids are repeated in this step until the centroids converge or a specified number of iterations is reached. After two more iterations the centroids of the clusters were no longer changing with their final values $C1(3.67, 9)$, $C2(7, 4.33)$, and $C3(1.5, 3.5)$. The final cluster assignments are as follows: cluster 1 includes the data points (2, 10), (5, 8), and (4, 9); cluster 2 includes the data points (4, 8), (5, 7), and (4, 6); and cluster 3 includes the data points (2, 5) and (1, 2) as shown in Table 2.4.

TABLE 2.4 Data points assignment to the clusters for second and third iterations

				C1	C2	C3	C1	C2	C3
			A1	4.5	7	1.67	3.67	7.00	1.50
	DATA POINTS		A2	8.5	4.3	5.67	9.00	4.33	3.50
NO.	A1	A2		DISTANCE 2			DISTANCE 3		
1	2	10		4.00	10.67	4.67	2.67	10.67	7.00
2	2	5		6.00	5.67	1.00	5.67	5.67	2.00
3	8	4		8.00	1.33	8.00	9.33	1.33	7.00
4	5	8		1.00	5.67	5.67	2.33	5.67	8.00
5	7	5		6.00	0.67	6.00	7.33	0.67	7.00
6	6	4		6.00	1.33	6.00	7.33	1.33	5.00
7	1	2		10.00	8.33	4.33	9.67	8.33	2.00
8	4	9		1.00	7.67	5.67	0.33	7.67	8.00

2.8 SUMMARY

This chapter equips readers with the requisite mathematical foundation for undertaking machine learning tasks. It guides learners through a structured progression, commencing with the fundamental mathematical concepts critical for comprehending machine learning principles. As readers progress, they develop the ability to mathematically represent machine learning models, fostering understanding and confident implementation. Furthermore, the chapter cultivates the essential skills of translating machine learning problems into mathematically optimized formulations. This empowers readers with problem-solving abilities in diverse machine learning contexts. The chapter also focuses on analyzing and interpreting mathematical expressions within machine learning algorithms, giving readers profound insights into the operational mechanisms driving these algorithms, ultimately enhancing their ability to leverage them effectively. Finally, the chapter equips learners to apply mathematical representations to evaluate algorithmic efficiency and model behavior.

Exercises

1. Assume that we have the following set of emails in Table 2.5 classified as either spam or ham. Given the new email *"review us now,"* find the probability that the given email (new email) is (i) Spam or (ii) Ham.

   TABLE 2.5 Email classification

EMAIL	LABEL
Send us your password	Spam
Send us your review	Ham
Password review	Ham
Review us	Spam
Send your password	Spam
Send your account	Spam

2. Three factories F1, F2, and F3 in the Dodoma region produce 50%, 25%, and 25%, respectively, of the total daily output of bottles of grape juice. It is known that 4% of the bottles of juice produced by Factories F1 and F2 are defective and that 5% of those produced in F3 are defective. If one bottle of juice is picked up at random from a day's production, calculate the probability that it is defective.
3. Suppose you are given the following set of data in Table 2.6 with the Boolean input variables a, b, and c, and a single Boolean output variable K.
   a. Assume we are using a naïve Bayes classifier to predict the value of K from the values of the other variables.

i. According to the naïve Bayes classifier, what is $P(K = 1 | a = 1 \wedge b = 1 \wedge c = 0)$?
ii. According to the naïve Bayes classifier, what is $P(K = 0 | a = 1 \wedge b = 1)$?

TABLE 2.6 Set of Boolean data

a	B	c	K
1	0	1	1
1	1	1	1
0	1	1	0
1	1	0	0
1	0	1	0
0	0	0	1
0	0	0	1
0	0	1	0

4. For the following scores of students in an examination: 84, 58, 90, 56, 85, 72, 64, 54, 48, 88, 92, and 74. Compute the:

   a. Measures of dispersion.
   b. Measures of central tendency.
   c. Quartiles.
   d. The 10th, 20th, 50th, and 70th percentiles.

5. Given the following data points.

X	Y
2	3
4	7
6	8
8	10
10	12

   Calculate the covariance between the predictor variable X and the response variable Y.

6. Given the following data points with two predictor variables X_1 and X_2 and one response variable Y.

X_1	X_2	Y
1	2	3
2	1	6
3	4	7
4	3	10
5	5	12

Calculate the covariance matrix between the predictor variables X_1, X_2, and the response variable Y.
7. Consider a neural network with an input $x = 2$, weight $w = 0.5$, and bias $b = 1$. Compute the output of the neuron using the hyperbolic tangent (tanh) activation function. Then, repeat the computation for an output neuron using the ReLU activation function with the output of the hidden neuron as its input and weight $w' = -1$, bias $c = 0.5$.
8. Using eigenvalues and eigenvectors for principal component analysis (PCA), perform dimensionality reduction on the following dataset.

$$\begin{pmatrix} 2 & 3 \\ 3 & 3 \\ 4 & 3 \end{pmatrix}$$

9. Perform K-Means clustering with $K = 2$ on the given dataset of points (2, 4), (1.5, 2), (3, 4), (1), (3, 2.5), and (1, 2), using your chosen initial centroids and the Euclidean distance method for distance calculation.
10. How does centroid initialization affect the K-means algorithm? Brainstorm strategies for centroid initialization and their implications.

FURTHER READING

Aggarwal, C. C. (2020). *Linear algebra and optimization for machine learning: A textbook.* Springer.
Alencar, M. S., & Alencar, R. T. (2024). *Set, measure, and probability theory.* CRC Press.
Bertsekas, D., & Tsitsiklis, J. N. (2008). *Introduction to probability* (Vol. 1). Athena Scientific.
Bhatia, P. (2019). *Data mining and data warehousing: Principles and practical techniques.* https://openlibrary.org/books/OL28937714M/Data_Mining_and_Data_Warehousing
Borovkov, A. A. (1999). *Probability theory.* CRC Press.
Bruce, P., Bruce, A., & Peter, G. (2020). *Practical statistics for data scientists* (Second Edition). O'Reilly Media, Inc.
Dalgaard, P. (2002). *Introductory statistics with R.* Springer.
Deisenroth, M. P., Faisal, A. A., & Ong, C. S. (2020). *Mathematics for machine learning.* Cambridge University Press.
Evans, M. J., & Rosenthal, J. S. (2004). *Probability and statistics: The science of uncertainty.* Macmillan.
Grinstead, C. M., & Snell, J. L. (1997). *Introduction to probability.* American Mathematical Society.
Haden, P. (2019). Descriptive statistics. In S. A. Fincher, & A. V. Robins (Eds.), *The Cambridge handbook of computing education research. Cambridge handbooks in psychology* (pp. 102–132). Cambridge University Press.

Hartigan, J. A., & Wong, M. A. (1979). Algorithm AS 136: A K-means clustering algorithm. *Applied Statistics/Journal of the Royal Statistical Society. Series C, Applied Statistics*, 28(1), 100. https://doi.org/10.2307/2346830

Mirkin, B. (2005). *Clustering for data mining: A data recovery approach.* https://ci.nii.ac.jp/ncid/BA71969362

Montgomery, D. C., & Runger, G. C. (2020). *Applied statistics and probability for engineers.* John Wiley & sons.

Ross, S. M. (2017). Introductory statistics. In Sheldon M. Ross (Ed.), *Introductory statistics* (Fourth Edition, pp. 797–800). Academic Press.

Strang, G. (2019). *Linear algebra and learning from data.* Wellesley-Cambridge Press.

Tabak, J. (2014). *Probability and statistics: The science of uncertainty.* Infobase Publishing.

Wasserman, L. (2004). *All of statistics.* Springer.

Data preparation 3

Upon completing this chapter, learners should be able to:

1. Understand the machine learning process.
2. Identify business problems that can potentially be solved using machine learning techniques.
3. Use different methods for collecting relevant data for machine learning tasks.
4. Apply various data preprocessing techniques to ensure quality and reliability.
5. Understand ethical considerations in data collection.

3.1 OVERVIEW OF MACHINE LEARNING PROCESS

Generally, the machine learning process entails several steps, such as understanding the problem to be addressed, collecting and preprocessing data for training the model, and evaluating and deploying the model. Such a process is depicted in Figure 3.1. This chapter focuses on the steps related to data preparation, including business problem identification, defining success criteria, and data collection and preprocessing. The remaining steps shown in Figure 3.1 are covered in Chapter 4.

3.2 BUSINESS PROBLEM IDENTIFICATION

A business problem is a specific challenge or issue an organization encounters in its day-to-day operations. It represents a gap between the current state of the business and its desired vision, hindering its performance or preventing it from achieving its goals. Business problems can vary in nature and complexity, requiring analysis, planning, and implementation of appropriate solutions to resolve them effectively.

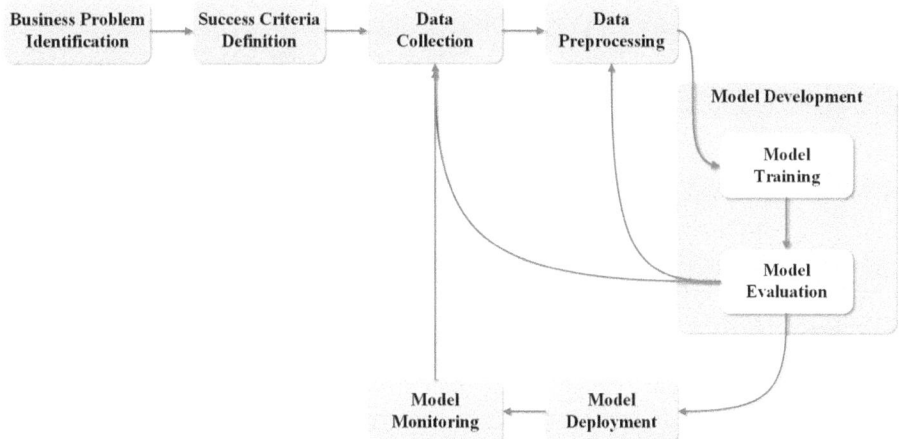

FIGURE 3.1 Machine learning process.

In the machine learning context, a business problem refers to a specific challenge or issue an organization faces that machine learning techniques can solve. In other words, a business problem is an opportunity that can benefit from leveraging data and machine learning techniques to make informed decisions, improve efficiency, and optimize processes that will ultimately lead to achieving business objectives. Identifying and clearly defining the business problem is a critical first step in the machine learning process, as it establishes the scope and direction for subsequent data collection, preprocessing, and modeling phases. The business problem may vary based on the nature of the problem domain. It could involve predicting customer behavior, disease diagnosis, fraud detection, weather forecasting, and recommendation systems.

3.3 SUCCESS CRITERIA DEFINITION

Success criteria refer to the specific benchmarks or goals established for a machine learning project. It defines what constitutes a successful outcome for the project and guides the evaluation of its progress and final results. They are typically defined in the business problem identification phase, where project objectives are identified and aligned with business goals. They also serve as a reference point when assessing whether the outcomes meet the desired requirements and provide business value. Some common examples of key success criteria include business objectives, measurable metrics (e.g., key performance indicators), timeframe, and stakeholder engagement.

3.4 DATA COLLECTION

Data is essential for accurately designing and implementing machine-learning models. Therefore, collecting the specific data related to the problem you intend to solve before embarking on a machine learning project is essential. Data can be gathered from pre-existing databases or can be built from scratch. Usually, the nature of the problem domain dictates how data should be collected and stored. For instance, specialized equipment is necessary to create a digital image catalogue when tasked with developing a system to identify skin cancer from skin images. In contrast, creating a recommendation system for e-commerce does not necessitate specialized data collection tools. Instead, all requisite data is supplied by users during product purchases. Notably, data collection process considerations include the nature of the data and their corresponding sources as detailed in the following sections.

3.4.1 Nature of data

Data comprises raw facts, figures, or statistics, which may exist in structured, semi-structured, or unstructured forms. Usually, data is represented in different formats such as numbers, text, images, audio, video, or any other format. Structured data is stored in a predefined format and is usually highly specific. A simple illustration of structured data is a Microsoft Excel file, in .xls or .csv format, where each column represents an attribute of the data. Unstructured data includes a multitude of diverse types of data typically stored in their native formats. A set of photo, video, or text files can represent unstructured data. Semi-structured data combines the features of unstructured and structured data. Examples of semi-structured data include JavaScript Object Notation (JSON), Extensible Markup Language (XML), and log files. Semi-structured data includes tags and elements, often called metadata, which serve to group the data and delineate its storage structure.

3.4.2 Data sources

Machine learning datasets can originate from available or online resources, or be built from primary sources. The online datasets may be either publicly accessible or proprietary. Therefore, utilizing these datasets demands thoroughly examining ethical considerations across different data lifecycle stages. This covers scrutiny in data sources and collection, data representation, and data balancing and splitting. It is imperative to uphold principles of fairness, transparency, and responsible data usage. Delving into the ethical dimensions at each stage is essential for fostering ethical practices in machine learning. Table 3.1 outlines a few online data sources where data for implementing machine learning models can be accessed.

Alternatively, the Google search engine can be used to search for datasets using relevant keywords and filter the results based on the dataset formats (e.g., images, text, and videos) or accessibility (i.e., freely available or not).

TABLE 3.1 Online dataset repositories

NAME OF DATASET REPOSITORY	DESCRIPTION
UCI Machine Learning Repository	The University of California Irvine (UCI) data repository provides free datasets for empirically analyzing machine learning models. The UCI repository can be accessed at https://archive.ics.uci.edu/ml/.
Kaggle	Regarded as one of the most resourceful data repositories and online communities that support the development of machine learning models. It is a rich repository, offering a vast and diverse collection of free datasets. Additionally, Kaggle has various tools for data exploration, visualization, and collaboration. It is a valuable platform for both beginners and experienced data scientists. The Kaggle repository is available at https://www.kaggle.com/datasets.
GitHub	Stores and publishes open machine learning datasets that are freely accessible for analyzing machine learning algorithms. The public datasets on GitHub can be accessed at https://github.com/awesomedata/awesome-public-datasets.
Microsoft Research Open Data	The repository contains free accessible datasets to promote research advancements in different fields, including computer vision, NLP, and domain-specific sciences. The repository is available at https://msropendata.com/.
OpenML	An online platform for machine learning that facilitates the sharing and organization of data, algorithms, and experiments. It aims to establish a seamless, interconnected ecosystem that integrates with existing processes, code, and environments. The platform enables global collaboration, allowing individuals to build upon each other's ideas, data, and results, regardless of the infrastructure and tools they use. The OpenML data repository is available at https://www.openml.org/.
Amazon Web Service (AWS) Datasets	Provides a lot of datasets for quick deployment of machine learning models when using AWS. Different third parties provide datasets under varied licenses that determine in which applications they can be used. The Amazon datasets repository is available at https://registry.opendata.aws/.
Zenodo Open Data Repository	An open-access platform that hosts a broad spectrum of research data across disciplines such as healthcare, agriculture, climate, and cyber security. With robust metadata standards and versioning capabilities, Zenodo facilitates collaboration and promotes transparency in scientific research. The Zenodo open data repository is available at https://zenodo.org/.
Hugging Face Dataset	An online platform for accessing and sharing datasets specifically suited for NLP, computer vision, and audio tasks. It also contains a variety of pre-trained models with the necessary tools for effectively using them. The Hugging Face open data repository is available at https://huggingface.co/datasets.

(*Continued*)

TABLE 3.1 (Continued)   Online dataset repositories

NAME OF DATASET REPOSITORY	DESCRIPTION
Government Open Data Portals	Are operated by governments, regional integration bodies, and international organizations thereby providing access to a wide range of datasets related to public services, the environment, demographics, and other topics. Examples of such portals are hosted by Tanzania, the United States of America, Canada, the European Union, and the World Bank at: https://www.nbs.go.tz/ https://www.data.gov/ https://open.canada.ca/en https://www.europeandataportal.eu/ https://data.worldbank.org/

3.4.3 Data curation

Data curation is essential when gathering information from multiple sources. This process involves collecting and standardizing data from diverse origins into a unified format. It entails employing relevant analysis tools and filtering methods to discern valuable data from irrelevant ones during integration. Typically, data curation tools aid in integrating, cleansing, adding metadata, validating, and preserving collected data. Ultimately, data curation enhances dataset accessibility and comprehension, making them more manageable for users to locate and interpret. Notable data curation tools include Alation, Talend, Stitch Data, Informatica, Ataccama ONE, and Alteryx. The choice of the exact tool depends on the properties and size of the data for a particular machine learning problem. A proper data curation process will ensure that the data remaining for labeling tasks are only those likely to enhance the performance of the models.

3.4.4 Data labeling

Data labeling is the process of identifying raw data and adding informative and meaningful labels to provide context for a machine learning algorithm to learn from. For instance, labels might indicate whether a photograph contains a dog or a cat, identify the words spoken in an audio recording, or specify whether an X-ray image shows a tumor. Data labeling is essential for several applications, such as image and text classification, action recognition, intrusion detection, and speech recognition. Figure 3.2 illustrates an example of labeled and unlabeled image samples. Notably, a labeled dataset from which an algorithm can learn is required for supervised learning. Typically, data labeling begins with the respective domain's experts (labelers or annotators) being asked to describe or group unlabeled pieces of data in their respective categories. For example, a medical domain expert may be requested to tag X-ray images based on the condition "Does the image contain signs of tuberculosis or not." Tagging can be a "yes" or "no" answer corresponding to whether a patient is infected with tuberculosis or not, respectively.

FIGURE 3.2 Labeled and unlabeled image samples.

3.4.5 Ethical considerations in data collection

Data often inherits societal biases that can be perpetuated by machine learning algorithms and impact outcomes, thereby reinforcing existing disparities and inequalities. Thus, it becomes imperative to conscientiously address ethical concerns throughout the data collection process, as highlighted in Table 3.2.

3.5 DATA PREPROCESSING

Usually real-world data typically contains noise, missing values, duplicate values, and outliers, and it may be in unusable format—making it unsuitable for directly developing machine learning models. Therefore, data preprocessing targets transforming raw data into a format appropriate for training machine learning algorithms. Data preprocessing can significantly affect the performance of a machine learning model. It entails critical steps, including data cleaning, transformation, dimensionality reduction, and integration, as described in the following subsections.

3.5.1 Data cleaning

Data cleaning deals with fixing missing, outlier, duplicate, corrupted, incorrectly formatted, and incorrect values within a dataset. Data with such issues could lead to unreliable machine learning models. Generally, data cleaning helps in reducing errors and improving data quality. Although the data cleaning process can be time-consuming and tedious, it should not be ignored. Several techniques can be used in data cleaning depending on the nature of the dataset, as described in the following subsections.

3.5.1.1 Removing duplicate or irrelevant values

Duplicate values in datasets often stem from different sources, such as data entry errors, merging data from multiple sources, or incomplete deletion of redundant records. Addressing duplicates is a critical aspect of the data-cleaning process. Failure to remove duplicates can lead to redundant information being fed into the model,

TABLE 3.2 Ethical considerations in data collection

ETHICAL ISSUE	DESCRIPTION
Privacy	Privacy concerns often stem from data containing personal and sensitive information, such as names, addresses, and financial details. Collecting and securely storing data is crucial to reduce the risk of unauthorized access. Individuals should also maintain control over their data usage.
Accuracy	Ensuring data accuracy requires rigorous validation and verification procedures to confirm precision and reliability. Thorough scrutiny and validation checks help prevent disseminating potentially misleading or inaccurate information.
Security	Employ encryption and access controls during data collection to restrict access to unauthorized personnel, mitigating the risk of unauthorized disclosure. Regularly audit data handling processes and comply with legal standards to promptly detect and address security vulnerabilities or breaches.
Ownership	Ethical data handling requires respecting individuals' rights to control their data and acknowledging their ownership. Organizations should establish clear policies on data ownership, outlining guidelines for control and usage to uphold ethical standards.
Transparency	Data transparency entails openly acknowledging biases, errors, or uncertainties within datasets, enabling informed decision-making and reducing potential harm. Embracing data transparency cultivates trust, accountability, and responsible data usage in machine learning applications.
Bias and Fairness	Data collection practices must avoid unfairly targeting or excluding specific groups, necessitating vigilance against potential biases in sampling and collection methods.
Informed Consent	Individuals whose data is collected should be informed about the purpose of the data collection, its intended use, and any potential benefits or risks. Besides, participants should also be allowed to decline participation or withdraw their consent at any time.
Accessibility	It entails removing barriers to data access, such as cost or technical expertise, and providing documentation and tools to facilitate understanding and utilization of the data. Prioritizing data accessibility promotes inclusivity, transparency, and collaboration, enabling broader participation and societal benefits from machine learning advancements.

resulting in wasted computational resources and skewed results. The typical approach to handling duplicates involves identifying and removing them, retaining only one unique observation for each duplicated entry. Similarly, irrelevant values not aligning with the problem at hand require attention. These values can be managed by deleting the corresponding observations or replacing the irrelevant ones with accurate ones, if available or retrievable.

3.5.1.2 Fixing structural errors

Structural errors occur due to typos, incorrect capitalization, or improper naming conventions. Such inconsistencies may lead to mislabeled categories or classes. For instance, you may find "N/A" and "Not Applicable" in a dataset, but they should be considered in the same category. In the case of structural error, the data (or entries) of the same category should be renamed using the same convention.

3.5.1.3 Detecting and removing outliers

Outliers refer to the data points in a dataset that are beyond a predefined distribution range and fall far from the mean of the dataset's observations. Usually, outliers appear not to fit within the dataset under analysis. Outliers could lead to unrealistic model performance and inflation of error metrics which give higher weight to large errors. Outliers can easily be detected using visualization techniques such as clustering, z-score, and box plots.

3.5.1.4 Handling missing values

Missing values are among the common challenges in datasets, occurring when certain attribute values are missing. Most machine learning algorithms cannot handle missing values, which may lead to errors or biased models if trained on such data. Failure to adequately address missing values can result in skewed models and prone to incorrect results. Missing values can be addressed using different approaches, such as:

- Dropping a feature or record with missing values. This is fairly simple but may lead to loss of information. Therefore, careful consideration, such as dataset size, is needed before dropping a feature or record.
- Filling missing values based on the measures of central tendency (mode, mean, and median). However, there is a risk of compromising data integrity due to working on assumptions rather than actual data.

3.5.1.5 Validation

Data validation involves inspecting data quality before training a machine learning algorithm. The following questions should be answered as part of data validation:

 i. Does the data make sense?
 ii. Does the data adhere to domain-specific rules?
 iii. Does the data support or refute your working theory or provide new insights?
 iv. Can you identify trends in the data to assist with developing your next theory?
 v. If not, is that because of issues in data quality?

3.5.2 Data Transformation

Data transformation entails converting data between formats, such as converting numerical data to categorical data through binning or categorical data to numerical data via encoding. Moreover, data transformation involves scaling the data in a suitable range through normalization. The following subsections describe common data transformation techniques.

3.5.2.1 Binning

Binning or discretization transforms numerical attributes into categorical equivalents. For instance, age values can be discretized into categories like 20–39, 40–59, and 60–79. Binning can enhance machine learning model accuracy by mitigating noise or non-linearity, aiding in outlier identification, and smoothing data through techniques like equal bin frequency, means, median, and boundaries.

3.5.2.2 Encoding

Machine learning algorithms operate solely on numerical data and cannot comprehend textual, date, or other non-numeric values. Encoding translates these diverse values into numerical formats, enabling algorithms to interpret and leverage them for learning and predictive tasks. Consequently, converting categorical values into numerical ones via encoding becomes imperative. Common encoding techniques in machine learning include Label Encoding, where each category receives a unique numerical label, and One-Hot Encoding, which generates binary columns representing the presence or absence of each category in the dataset. For instance, when applying One-Hot Encoding to a binary attribute like gender, with male or female values, the resulting encoded values become zero (0) or one (1), respectively, indicating male or female. Notably, encoding methods are important in the preprocessing stage before inputting data into machine learning algorithms. This ensures efficient interpretation of diverse information types within the dataset, thus deriving meaningful patterns.

3.5.2.3 Data normalization

Data normalization refers to changing the numerical values of attributes to a common scale without affecting the differences or losing information. Normalization provides equal weights or importance to each attribute so that no single attribute influences the performance of a model because of its large values. For example, a dataset can have several attributes with values in the order of tens and others in the order of millions. In this case, normalization will scale down all attributes to a common scale (say 0 to 1). This process is also known as rescaling attribute values. This technique is particularly useful for algorithms that rely on distance measures, such as k-NN. The most widely used technique is min-max normalization, which performs a linear transformation of the original data to fit it in the range of 0 to 1. By so doing, it ensures that all attributes are handled equally regardless of their original values. It is computed by subtracting the minimum value from each feature and dividing the result by the range (maximum-minimum) as expressed in Equation (3.1).

$$normalized\_value = \frac{(feature\_value - min\_value)}{(max\_value - min\_value)} \qquad (3.1)$$

3.5.2.4 Standardization

Standardization transforms numerical data to have zero mean and unit standard deviation. Unlike normalization, which scales the data within a specific range, standardization focuses on centering the data around the mean and adjusting its distribution. The most commonly used standardization technique is z-score, which transforms features from differing means and standard deviations to a standard Gaussian distribution. Z-score is the most suitable technique when there are outliers in the dataset. The z-score standardization formula is expressed in Equation (3.2).

$$\text{standardize\_value} = \frac{(\text{feature\_value} - \text{mean})}{\text{standard\_deviation}} \quad (3.2)$$

Tables 3.5 and 3.6 show examples of features before and after z-score standardization, respectively.

TABLE 3.3 Features before normalization

INDEX	PREGNANT	GLUCOSE	BP	SKIN	INSULIN	BMI	PEDIGREE	AGE
0	6	148	72	35	0	33.6	0.627	50
1	1	85	66	29	0	26.6	0.351	31
2	8	183	64	0	0	23.3	0.672	32

TABLE 3.4 Features after normalization

INDEX	PREGNANT	GLUCOSE	BP	SKIN	INSULIN	BMI	PEDIGREE	AGE
0	0.353	0.744	0.590	0.354	0.000	0.501	0.234	0.483
1	0.059	0.427	0.541	0.293	0.000	0.396	0.117	0.167
2	0.471	0.920	0.525	0.000	0.000	0.347	0.254	0.183

TABLE 3.5 Features before z-score standardization

INDEX	PREGNANCIES	GLUCOSE	BLOOD PRESSURE	SKIN THICKNESS	INSULIN	BMI	DIABETES PEDIGREE FUNCTION	AGE
0	6	148	72	35	0	33.6	0.627	50
1	1	85	66	29	0	26.6	0.351	31
2	8	183	64	0	0	23.3	0.672	32

TABLE 3.6 Features after z-score standardization

INDEX	PREGNANCIES	GLUCOSE	BLOOD PRESSURE	SKIN THICKNESS	INSULIN	BMI	DIABETES PEDIGREE FUNCTION	AGE
0	0.639947	0.848324	0.149641	0.907270	−0.692891	0.204013	0.468492	1.425995
1	−0.844885	−1.123396	−0.160546	0.530902	−0.692891	−0.684422	−0.365061	−0.190672
2	1.233880	1.943724	−0.263941	−1.288212	−0.692891	−1.103255	0.604397	−0.105584

3.5.3 Exploratory data analysis

Exploratory Data Analysis (EDA) utilizes statistical summaries and graphical representations to analyze data, aiming to uncover trends and patterns or validate assumptions. Its primary goal is to extract meaning from the data and glean insights before constructing a machine learning model. EDA goes beyond mere numerical analysis, it delves into understanding the narrative within the data, unveiling patterns, and fostering a profound comprehension of the dataset before it is used in machine learning algorithms.

EDA typically begins with a descriptive overview of the dataset, encompassing checks on its dimensions (number of columns and rows), comprehension of feature data types, and identification of missing values. Visualizations such as box plots, histograms, and scatter plots serve the purpose of investigating distributions, central tendencies, and potential outliers within numerical data. Furthermore, EDA involves the analysis of relationships between variables, utilizing correlation matrices or pair plots to discern associations among features. Bar charts or frequency tables come into play for comprehending distributions across various categories in categorical data. The key methodologies of EDA are elaborated in the subsequent subsections.

3.5.3.1 Data summarization

Data summarization provides a summary or report of data in an informative and understandable manner. The summary contains some necessary statistical explanations about the data, such as the minimum and maximum value of the feature across all entries. For instance, in Table 3.7, each feature column has a summary that shows statistical explanations of the data, such as count, mean, standard deviation, variance, percentiles, and interquartile range. The summary helps to show whether the values of the features are informative and comprehensible.

3.5.3.2 Data visualization

Data visualization is transforming data into a visual or graphical format (such as graphs, maps, and charts) so that it can be easily understood and communicate insights from data to a wide audience. Data visualization is essential as it identifies patterns, trends,

TABLE 3.7 Data summarization example

	PREGNANT	GLUCOSE	BP	SKIN	INSULIN
count	768	768	768	768	768
mean	3.845052	120.894531	69.105469	20.536458	79.799479
std	3.369578	31.972618	19.355807	20.536458	115.244002
min	0.000000	0.000000	0.000000	0.000000	0.000000
25%	1.000000	99.000000	62.000000	0.000000	0.000000
50%	3.000000	117.000000	72.000000	23.000000	30.500000
75%	6.000000	140.250000	80.000000	32.000000	127.250000
max	17.000000	199.000000	122.000000	99.000000	846.000000

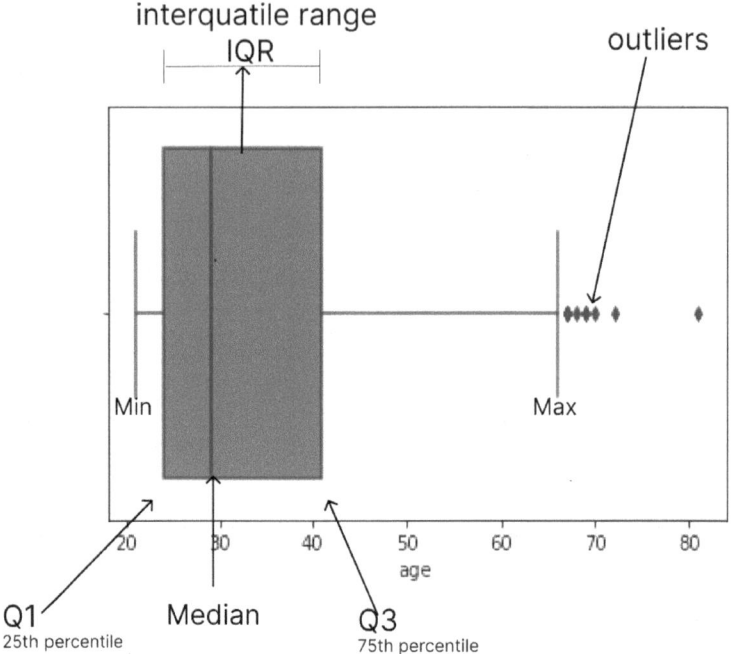

FIGURE 3.3 Data distribution of a single feature.

outliers, and variable distributions. It also aids in identifying data quality issues, such as inconsistencies, errors, or missing values, before the data preprocessing stage. It is particularly valuable for individuals who may lack technical aspects of the data. By visually representing the data, complex information becomes more accessible, facilitating a better understanding of the dataset and aiding in the effectiveness of data preprocessing. Figure 3.3 depicts an example of data distribution of a single feature (i.e., age) plotted individually for distribution analysis.

3.5.4 Types of exploratory data analysis

There are three types of EDA: univariate, bivariate, and multivariate.

3.5.4.1 Univariate

In univariate analysis, one feature (numerical or categorical) is analyzed independently and in detail. The feature is analyzed to observe and learn its distribution and central measure of tendency values such as mean, mode, and median to gain insight into the data. The feature can also be visualized with the help of graphical tools for easier interpretations. Graphs to visualize a single feature can be pie charts, bar plots, and histograms, as shown in Figure 3.4.

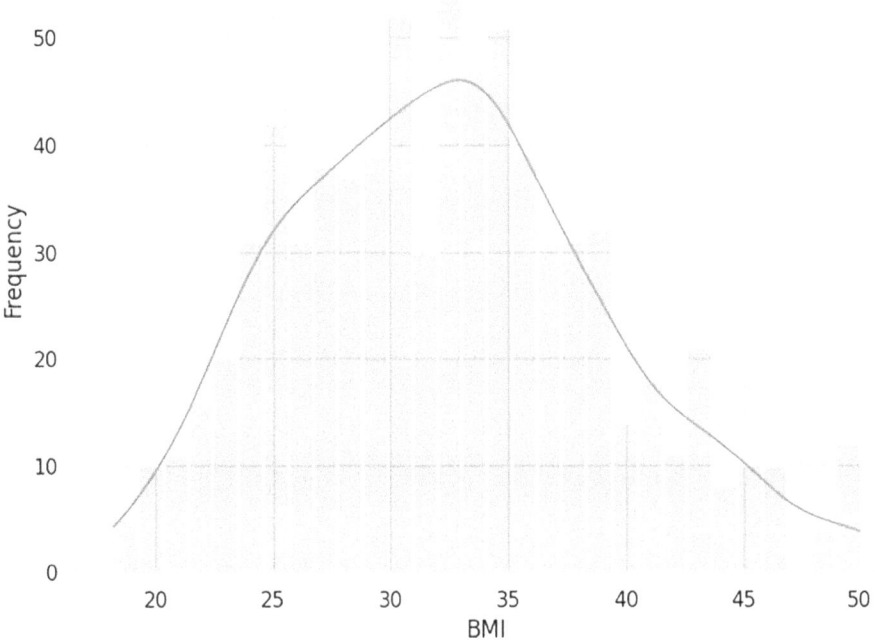

FIGURE 3.4 Univariate analysis example.

3.5.4.2 Bivariate

Bivariate analysis involves the analysis of two independent attributes simultaneously. The features involved can be numerical, categorical, or any combination of both. The analysis aims to discover the relationship between the two attributes if there is a difference or association between them. The features are visualized in the same plot graph to learn their relationship, as shown in Figure 3.5. The two features can be visually analyzed by using any of the following approaches:

- Scatterplots and heatmaps (for numerical and numerical attributes).
- Stacked column chart, Chi-square test, and Combination chart (for categorical and categorical attributes).
- Line chart with error bars, z-test, t-test, and combination chart (for categorical and numerical attributes).

3.5.5 Multivariate

Multivariate analysis is crucial when analyzing more than two independent features simultaneously, as depicted in Figure 3.6. Multivariate analysis includes various techniques, such as cluster analysis, factor analysis, multiple regression analysis, and

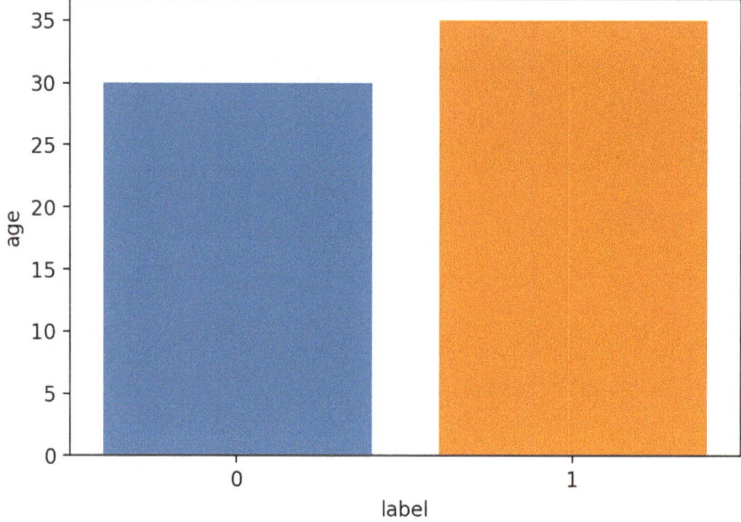

FIGURE 3.5 Bivariate analysis example.

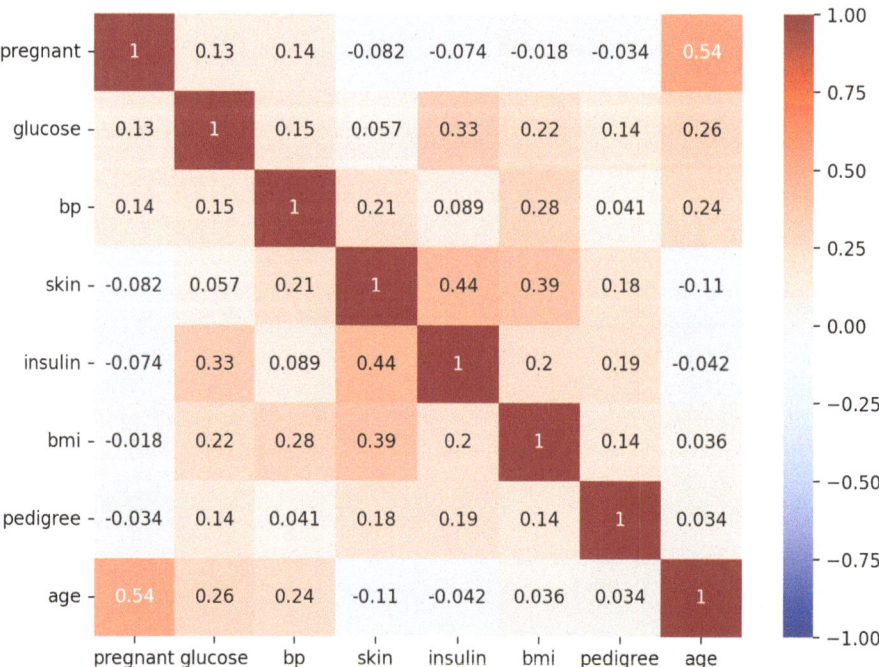

FIGURE 3.6 Multivariate analysis example.

principal component analysis (PCA), among others. Such methods facilitate a comprehensive exploration of complex relationships and patterns across the features, catering to different characteristics of the dataset. In contrast to univariate analysis, which focuses on one variable at a time, multivariate analysis considers the dependencies and interactions between multiple variables. Multivariate analysis enables a deeper understanding of the underlying structure and dynamics of the data.

3.5.6 Dimensionality reduction

High-dimensional datasets are often challenging to visualize and comprehend. Therefore, dimensionality reduction is usually applied to convert a dataset from a higher-dimensional space to a lower-dimensional one while preserving its original information. This technique is utilized when a dataset comprises many input features. Therefore, the goal is to eliminate the less important features and avoid complicating the modeling task. Dimensionality reduction is commonly applied in domains involving high-dimensional data, for example, signal processing, speech recognition, and bioinformatics. The following subsections briefly highlight the common dimensionality reduction techniques.

3.5.6.1 Feature selection

Feature selection is a process of automatically selecting informative features that have the most significant impact on the performance of a machine learning model. Having irrelevant features in the dataset can reduce the performance of machine learning models, especially linear algorithms like simple linear and logistic regression. The common benefits of feature selection include the following:

- **Reduces Overfitting**: Feature selection reduces model overfitting by identifying and using only the most relevant features for model training, discarding redundant or irrelevant ones. Ultimately improving the performance of the model.
- **Reduces Training Time**: Fewer features mean that models train faster.

Notably, backward and forward feature elimination methods are the common techniques used to perform feature selection, as detailed in the following:

3.5.6.1.1 Backward feature elimination
This technique is employed to systematically remove features that exhibit minimal impact on predicting the output or dependent feature. It commences with a full set of features and progressively eliminates the least influential ones until a specified stopping point is reached. This iterative process rigorously refines the feature set, enhancing the model's efficiency and interpretability. This method ensures the model focuses solely on the most impactful features, thereby refining predictive accuracy and streamlining the overall model complexity.

3.5.6.1.2 Forward feature selection
This technique is the inverse of backward feature elimination. In this approach, features are not removed but progressively added based on their ability to enhance the model's performance. This method systematically evaluates and selects features that can effectively improve the model's predictive accuracy, in other words, prioritizing those that yield the highest increase in performance. Generally, the model refines its understanding by iteratively including the most influential features, ensuring a more robust and optimized configuration that bolsters its predictive capabilities.

3.5.6.2 Feature extraction

Feature extraction involves selecting or transforming the most relevant and informative features from raw data, streamlining it for more effective model training. This process identifies key patterns or attributes within the data that contribute significantly to the task at hand, enhancing decision-making and predictive accuracy in machine learning tasks. For instance, text analysis may entail converting words into numerical representations or pinpointing important phrases that convey a sentence's meaning. In image processing, it could involve recognizing edges, textures, or shapes that distinguish one object from another. Eliminating redundant or less important information aids in focusing on the most crucial aspects that improve the model's performance. This streamlined data enhances the ability of the model to identify essential patterns, leading to more correct predictions and improved decision-making in machine learning applications.

The common technique for feature extraction is Principal Component Analysis (PCA). The PCA is a statistical technique that transforms correlated features into a set of linearly uncorrelated features through orthogonal transformation. The resultant features, known as principal components, capture the essential information in the data while reducing its dimensionality. PCA evaluates the variance of each feature, prioritizing those with high variance to retain valuable information and enhance interpretability. Real-world applications span diverse domains such as movie recommendation systems, image processing, and optimizing power allocation in communication channels. PCA can inadvertently amplify existing biases in the data, potentially resulting in unfair outcomes if the data is skewed. Therefore, carefully selecting principal components is crucial to avoid excluding pertinent information and ensure fair and unbiased classifications.

3.5.7 Data balancing

Data imbalance is a common issue in machine learning, where one class or category within a dataset has significantly more representation than others. This can occur naturally, such as in fraud detection, where fraudulent transactions are far less frequent than legitimate ones, or due to biases in data collection. Uncorrected imbalances can lead to models that are heavily biased toward the majority class, thereby underperforming when encountering samples of the minority classes. Data balancing is a crucial technique that involves adjusting the distribution of classes to create a more balanced dataset. This might be achieved through oversampling (replicating minority class samples), undersampling (removing majority class samples), or more sophisticated approaches like the

Synthetic Minority Oversampling Technique (SMOTE). It is important to note that data balancing might not be necessary in all cases. Factors such as the severity of imbalance and project goals dictate its importance.

3.6 SUMMARY

This chapter provided the key steps in business problem identification, data collection, and preprocessing in machine learning. It began by underlining the importance of aligning machine learning initiatives with business goals, emphasizing the need to contextualize and define problems within the broader organizational landscape. Furthermore, the chapter explored the nature of data, highlighting various data sources and their essential characteristics. Subsequently, the chapter focused on data curation, cleaning, and labeling, outlining essential procedures to ensure data accuracy and coherence. It also discussed techniques for managing missing values and eliminating duplicates, thereby enhancing the integrity of the dataset. Moreover, the chapter introduced methods for data transformation, normalization, and exploratory data analysis (EDA) to uncover insights into data patterns and relationships. Finally, it introduced methods for dimensionality reduction, feature selection, and the utilization of principal component analysis (PCA) to streamline data preprocessing for enhanced model performance.

Exercises

1. Formulate a hypothetical business problem where machine learning can offer significant value. Describe the problem context, its alignment with business goals, and potential machine learning applications.
2. Research and compile a list of diverse data sources applicable to weather forecasting. Discuss the types of data available, their relevance, and the challenges associated with integrating multiple sources for machine learning models.
3. Devise a comprehensive data collection plan for a healthcare analytics project centered on patient outcomes. Outline data collection methodologies, anticipated challenges, and potential strategies to overcome them.
4. Find a dataset with missing and duplicate values from the data repository introduced in this chapter and implement data-cleaning techniques to rectify these issues. Document the steps taken and justify the chosen methods for cleaning the dataset.
5. Choose any dataset from the data repositories introduced in this chapter, apply data transformation techniques like normalization or scaling, and provide visual representations of the data through exploratory data analysis (EDA) methods. Interpret any observed trends or patterns.

6. When applying dimensionality reduction methods such as PCA to a dataset with high dimensions, what are its impacts on data representation and computational efficiency?
7. Choose any dataset with several features from the data repositories introduced in this chapter, and use feature selection methods to find the most impactful features for model development. Justify your selection criteria.
8. Analyze possible challenges one may encounter during data collection from highly specialized domains (e.g., healthcare and autonomous vehicles) and propose strategies to address them.
9. A financial institution uses historical loan data to train a machine learning model for loan approvals. Describe potential biases that may manifest in this dataset. Outline practical strategies to identify and mitigate such biases before and during model development.
10. Discuss the role of dimensionality reduction in preventing model overfitting.

FURTHER READING

Barga, Roger, Fontama, V., Wee Hyong Tok, Barga, R., Fontama, V., and Wee Hyong Tok. (2015). Data preparation. In *Predictive Analytics with Microsoft Azure Machine Learning*, 45–79.
Berman, Jules J. (2018). *Principles and practice of big data: preparing, sharing, and analyzing complex information*. Academic Press.
Bowles, Michael. (2015). *Machine learning in Python: essential techniques for predictive analysis*. John Wiley & Sons.
Brownlee, Jason. (2020). Data preparation for machine learning: data cleaning, feature selection, and data transforms in Python. *Machine Learning Mastery*.
Cielen, Davy, & Meysman, Arno. (2016). *Introducing data science: big data, machine learning, and more, using Python tools*. Simon and Schuster.
Dangeti, Pratap (2017). *Statistics for machine learning*. Packt Publishing Ltd.
Flach, Peter. (2012). *Machine learning: the art and science of algorithms that make sense of data*. Cambridge University Press.
Kelleher, John D., Namee, Brian Mac, & D'Arcy, Aoife. (2020). *Fundamentals of machine learning for predictive data analytics: algorithms, worked examples, and case studies*. MIT press.
Kononenko, Igor, & Kukar, Matjaz. (2007). *Machine learning and data mining*. Horwood Publishing.
Pyle, Dorian. (1999). *Data preparation for data mining*. Morgan Kaufmann.

Machine learning operations

4

Upon completing this chapter, learners should be able to:

1. Choose a suitable algorithm depending on the problem at hand and the nature of the data.
2. Explain the key steps for developing a machine learning model.
3. Describe the concepts of overfitting and underfitting and the strategies to mitigate them.
4. Apply optimization techniques for machine learning algorithms to enhance model performance.
5. Explain the key steps for deploying and monitoring machine learning models to ensure continued performance.

4.1 MODEL DEVELOPMENT

This chapter focuses on the remaining steps of the machine learning process, as depicted in Figure 3.1. The first step in machine learning operations is model development, which entails training and evaluation. Before developing the model, it is necessary to perform data splitting and select a specific algorithm, as discussed in the following subsections.

4.1.1 Dataset splitting

Dataset splitting involves dividing the dataset into training and testing sets. The training set is used to train the model, whereas the testing set is used to assess the performance of the model based on data it has not seen before. The rationale for using a testing set is to avoid assessing a model's performance based on seen (training) data, which could lead to unrealistic results. Dataset splitting can be done in two ways: Hold-Out and Cross-Validation.

4.1.1.1 Hold-out

Hold-out refers to reserving a subset of the dataset for testing while using the remainder for training machine learning models. Typically, a dataset is split in a specific percentage, for example, 70 by 30 or 90 by 10, where the larger segment set is allocated for training and the smaller segment set for testing. Usually, the training set is recommended to be in the range of 70% to 90% of the whole dataset.

4.1.1.2 Cross-validation

Cross-validation involves splitting a dataset into k subsets of equal size called folds. The model undergoes iterative training on $k-1$ folds while being tested on the remaining fold, ensuring that each subset serves both training and testing purposes. This helps to assess the model's ability to generalize to new, unseen data. Note that a k-fold cross-validation technique helps achieve an unbiased estimate of the model's performance when only a limited amount of data is available. Suppose the dataset is split into five equal subsets, as shown in Figure 4.1, forming a fivefold cross-validation. This implies that the model will

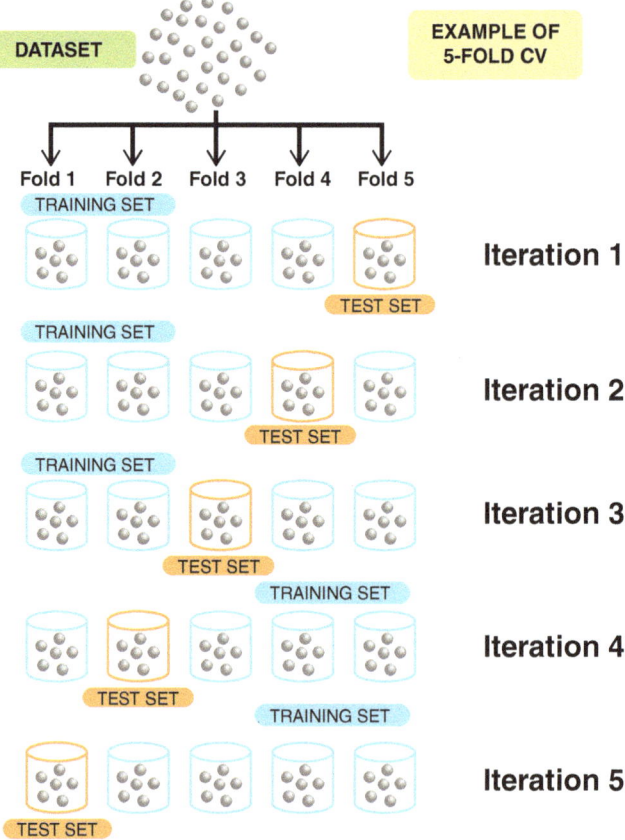

FIGURE 4.1 5-fold Cross-validation example.

be trained in five iterations. In each iteration, the model undergoes training utilizing four of the five subsets, while the remaining subset is used as the testing subset.

4.1.2 Choosing an algorithm

Before building machine learning models, it is essential to determine appropriate algorithms that align with the problem at hand and the available dataset. Determining the suitable machine-learning algorithm depends on various factors such as the problem at hand, algorithm capabilities, and the computational resources as described in the following subsections.

4.1.2.1 Problem understanding

A thorough understanding of the problem to be solved is crucial. This includes identifying whether the problem is classification, regression, clustering, or association rule mining. After identifying the type of problem, multiple machine learning algorithms within that specific problem category, as presented in Chapter 1, are trained to build models. Consequently, the model that exhibits the highest performance is chosen as the most suitable solution for the identified problem.

4.1.2.2 Algorithm capabilities

Each algorithm possesses unique strengths and weaknesses. For instance, decision trees excel in interpretability, making them valuable for understanding the underlying logic of a model, whereas neural networks are effective at addressing complex patterns within data.

4.1.2.3 Computational resources

Some algorithms might demand substantial computational resources, particularly when handling large datasets. Consider the computational complexity of each algorithm and select the one that performs better with the available resources.

4.1.3 Model training

Model training enables the selected algorithms to extract knowledge from the provided dataset. It is a critical step where a model progressively enhances its capability to predict the given data samples. Typically, the dataset undergoes splitting into training and testing sets, as described earlier in this chapter. Subsequently, the machine learning model engages with the training set, iteratively refining its performance by recognizing patterns and making predictions. This involves adjusting the algorithm's internal parameters, often represented as coefficients in a mathematical function, to capture the underlying patterns in the dataset better. The model refines its capacity to make accurate predictions for new, unseen data samples through this iterative process.

4.1.4 Model evaluation

Model evaluation entails evaluating the performance and effectiveness of a trained model on unseen data from the testing set. It is a crucial step for determining the ability of the trained model to generalize to new data and whether it meets the desired objectives of the problem. The primary purpose of using a testing set is to reveal the performance of the model on real-world data to ensure its reliability and effectiveness in practical applications. Several evaluation metrics are used to measure the performance of the trained model, depending on the nature of the problem. Table 4.1 presents the commonly used evaluation metrics for classification, regression, clustering, and association rule problems.

The mathematical presentations of the evaluation metrics shown in Table 4.1 are highlighted in the following formulas. Some of these formulas are derived from a fundamental tool known as the confusion matrix, presented in Table 4.2. This matrix captures the model's prediction results by comparing them with the actual labels in the dataset. At its core, the confusion matrix breaks down the classification results into four distinct categories:

- **True Positives (TP)**: This happens when the outcome is correctly predicted as positive when it is indeed positive. For example, a spam email is correctly predicted as spam.

TABLE 4.1 Performance metrics

PERFORMANCE METRICS	PROBLEM TYPE
Accuracy	Binary and multiclass classification.
Precision (Positive Predictive Value)	Binary and multiclass classification.
Recall (Sensitivity, True Positive Rate)	Binary and multiclass classification.
F1 Score	Binary and multiclass classification.
Area under the Receiver Operating Characteristic curve (AUC-ROC)	Binary classification.
Log Loss (Cross-Entropy Loss)	Binary and multiclass classification.
Mean Absolute Error (MAE)	Regression
Mean Squared Error (MSE)	Regression
Root Mean Squared Error (RMSE)	Regression
R-squared (Coefficient of Determination)	Regression
Silhouette Score	Clustering
Support, Confidence, and Lift	Association Rules Mining

TABLE 4.2 Confusion matrix

	ACTUAL POSITIVE	*ACTUAL NEGATIVE*
Predicted positive	True Positive (TP)	False Positive (FP)
Predicted negative	False Negative (FN)	True Negative (TN)

- **True Negatives (TN)**: This happens when the outcome is correctly predicted as negative when it is indeed negative. For example, a non-spam email is correctly predicted as non-spam.
- **False Positives (FP)**: This happens when the outcome is wrongly predicted as positive when it is indeed negative. This is also known as the Type 1 error. For example, a non-spam email is wrongly predicted as spam.
- **False Negatives (FN)**: This happens when the outcome is wrongly predicted as negative when it is indeed positive. This is also known as the Type 2 error. For example, a spam email is incorrectly predicted as non-spam.

Accuracy: This performance metric quantifies the proportion of correctly classified instances to the total number of instances evaluated. Accuracy is calculated as shown in Equation (4.1).

$$\text{Accuracy} = \frac{TP + TN}{TP + FP + FN + TN} \quad (4.1)$$

High accuracy indicates the model's ability to make correct predictions, whereas low accuracy suggests a higher rate of incorrect predictions.

Precision (i.e., Positive Predictive Value): This performance metric quantifies the proportion of correctly predicted positive instances among all instances predicted as positive, as given by Equation (4.2).

$$\text{Precision} = \frac{TP}{TP + FP} \quad (4.2)$$

A high precision value signifies that the model has a low rate of false positives, making it more reliable in its positive predictions.

Recall (i.e., Sensitivity or True Positive Rate): This performance metric quantifies the proportion of true positive instances correctly predicted by the model among all actual positive instances as given by Equation (4.3).

$$\text{Recall} = \frac{TP}{TP + FN} \quad (4.3)$$

A high recall value signifies that the model effectively captures a large proportion of positive instances.

F1 Score (i.e., F-Measure): This performance metric is the harmonic mean of precision and recall, providing a balanced assessment of the performance of the model on both positive and negative instances, as given in Equation (4.4). It is particularly useful in a scenario where the dataset has a disproportionate distribution of classes (i.e., it is an imbalanced dataset), as it prevents the evaluation from being overly influenced by the majority class.

$$\text{F1 Score} = 2 \times \frac{\text{Precision} \times \text{Recall}}{\text{Precision} + \text{Recall}} \quad (4.4)$$

A high F1 score indicates the model's strong ability to balance precision and recall. In contrast, a low F1 score suggests that the model struggles to achieve a balance between precision and recall, possibly favoring one over the other.

Specificity: This performance metric measures the proportion of true negative instances correctly predicted by the model among all actual negative instances, as given by Equation (4.5).

$$\text{Specificity} = \frac{TN}{TN + FP} \tag{4.5}$$

A high specificity value demonstrates that the model can capture a large proportion of negative instances.

Area under the Receiver Operating Characteristic Curve (AUC-ROC): This performance metric visually illustrates the balance between the true positive rate (i.e., Sensitivity) and the false positive rate (i.e., 1—Specificity) across different thresholds for the model. The ROC curve (depicted in Figure 4.2) is generated by plotting the true positive rate (TPR) against the false positive rate (FPR) across various classification thresholds. The graph's diagonal line ($y = x$) serves as a reference for random guessing.

The AUC-ROC value closest to the upper left corner signifies strong model performance in distinguishing between positive and negative instances. When comparing two ROC curves, the higher and more toward the upper left corner represents the superior model. AUC-ROC values near 0.5 (at $y = x$) suggest performance equivalent to random chance, whereas values below 0.5 indicate a model is performing worse than random guessing and potentially inverting predictions.

Log Loss (i.e., Cross-Entropy Loss or Logistic Loss): The log loss metric evaluates a model's performance when it assigns probability scores to various classes. It

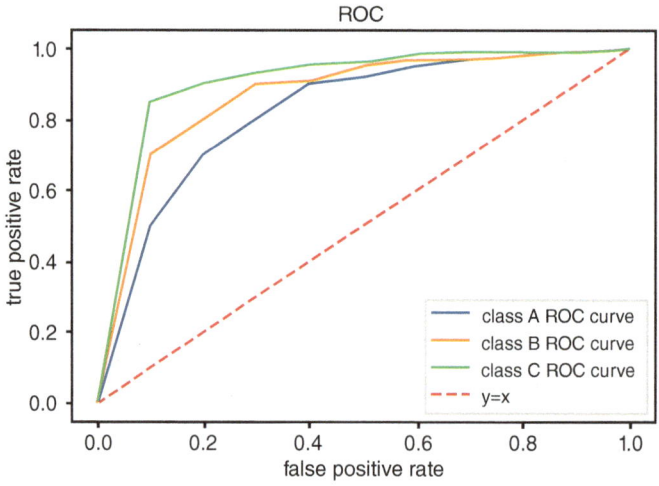

FIGURE 4.2 ROC Curves.

quantifies the disparity between the true label distribution and the predicted probabilities assigned by the model. Log loss is computed as given in Equation 4.6.

$$\log \text{Loss} = -\frac{1}{N}\sum_{i=1}^{N}\left(y_i \cdot \log(p_i) + (1 - y_i) \cdot (\log(1 - p_i))\right) \qquad (4.6)$$

where:

- N is the number of instances in the dataset.
- y_i is the true label for instance ii (0 or 1).
- p_i is the predicted probability that instance i belongs to class 1.

Mean Absolute Error (MAE): This performance metric quantifies the average absolute differences between predicted and actual values, providing a straightforward and interpretable measure of the model's accuracy. It computes the average absolute deviations of predictions from the true values, as shown in Equation (4.7).

$$\text{MAE} = \frac{1}{N}\sum_{i=1}^{N}\left|\text{predicted}_i - \text{actual}_i\right| \qquad (4.7)$$

where:

- N is the number of instances in the dataset.
- predicted_i is the predicted value for instance i.
- actual_i is the true value for instance i.

MAE values vary from 0 to ∞, with lower values signifying higher model performance. An MAE of zero indicates a perfect model, with predictions that exactly match the actual data. MAE is also widely employed in cases where anticipating the precise numeric value is critical, such as finance, where pricing must be predicted, or demand forecasting.

Mean Squared Error (MSE): This performance metric is used to quantify the average squared difference between predicted and actual values. It quantifies the overall accuracy of a regression model by averaging the squared errors across all instances in the dataset. The advantage of MSE over MAE lies in its ability to provide greater sensitivity to larger errors and deviations from true values, facilitating better optimization and model tuning. MSE is calculated as shown in Equation (4.8).

$$\text{MSE} = \frac{1}{N}\sum_{i=1}^{N}\left(\text{predicted}_i - \text{actual}_i\right)^2 \qquad (4.8)$$

where:

- N is the number of instances in the dataset.
- predicted_i is the predicted value for instance i.
- actual_i is the true value for instance i.

Root Mean Squared Error (RMSE): This performance metric is widely used to quantify the average magnitude of the errors between predicted and actual values. It is similar to MSE, but RMSE addresses one of the limitations of MSE by taking the square root of the average squared differences. This results in a quantity that is in the same units as the target variable, making it more interpretable. RMSE is calculated as shown in Equation (4.9).

$$\text{RMSE} = \sqrt{\frac{1}{N}\sum_{i=1}^{N}\left(\text{predicted}_i - \text{actual}_i\right)^2} \tag{4.9}$$

where:

- N is the number of instances in the dataset.
- predicted_i is the predicted value for instance i.
- actual_i is the true value for instance i.

R-squared (i.e., Coefficient of Determination): This performance metric assesses the model's goodness of fit by indicating the extent to which the independent variables elucidate the variability in the dependent variable. It is calculated as shown in Equation (4.10).

$$R^2 = 1 - \frac{\text{Sum of Squared Residuals}\,(\text{SSR})}{\text{Total Sum of Squares}\,(\text{TSS})} = \frac{\text{RSS}}{\text{TSS}} = 1 - \frac{\text{ESS}}{\text{TSS}} \tag{4.10}$$

where:

- RSS is the residuals or regression sum of squares. It measures the difference between the predicted and mean values of the dependent variable.
- TSS is the total sum of squares. It measures the difference between the actual and the mean values of the dependent variable.
- ESS is the error sum of squares. It measures the difference between the predicted and actual values of the dependent variable.

The R-squared values vary between 0 and 1, representing the extent to which the model explains the variance between dependent and independent variables. A higher value signifies a stronger model fit and better predictive performance for the dependent variable, while a lower value indicates limitations in the ability of the model to predict the dependent variable.

Silhouette Score: This performance metric measures the resemblance of a sample to its assigned cluster (cohesion) compared to other clusters (separation). Silhouette Score values range from -1 to $+1$, where a score close to $+1$ suggests well-clustered data points, a score close to 0 indicates an overlapping cluster or a cluster with ambiguous boundaries, and a score close to -1 suggests potential misassignment of data points to the wrong clusters. While the Silhouette Score is useful, it should be supplemented with other validation methods, particularly in scenarios with irregularly shaped or differently sized clusters. Silhouette Score is calculated as shown in Equation 4.11.

$$S_i = \frac{b_i - a_i}{\max(a_i, b_i)} \qquad (4.11)$$

where:

- a_i represents the average distance from the ith data point to other data points within the same cluster.
- b_i represents the average distance from the ith data point to data points in a different cluster, minimized across all clusters.

Support: This performance metric measures how often an itemset appears in a transaction. It is computed as the ratio of the number of transactions containing the itemset by the total number of transactions, as shown in Equation (4.12). The support value ranges from 0 to 1, where 0 indicates that the itemset does not appear in any transaction, and 1 indicates that the itemset appears in every transaction. Intermediate values between 0 and 1 represent the proportion of transactions in which the itemset appears. The higher the value of support the greater the prevalence and importance of the itemset in the transaction.

$$\text{Support}(X) = \frac{\text{Transactions containing } X}{\text{Total Transactions}} \qquad (4.12)$$

where:

- X represents itemsets.

Confidence: This performance metric measures the likelihood that the item will be present in a transaction given the presence of another related item in that transaction. Mathematically, confidence is defined as the ratio of the number of transactions containing both items X and Y to the number of transactions containing item X as shown in Equation (4.13). The confidence value ranges between 0 to 1. A value approaching 1 indicates a strong association, suggesting that the occurrence of Y is highly likely when X is observed. In contrast, a value close to 0 signifies a weaker connection, implying that the presence of X provides less certainty about the occurrence of Y in a transaction.

$$\text{Confidence}(X \rightarrow Y) = \frac{\text{Support}(X, Y)}{\text{Support}(X)} \qquad (4.13)$$

where:

- $X \rightarrow Y$ represents the association rule where a transaction containing item X also contains item Y.

Lift: This metric quantifies the strength of association between two items beyond what would be expected by chance. It compares the likelihood of the items occurring together in transactions to the likelihood of the items occurring independently of each other. Mathematically, Lift is computed as shown in Equation 4.14.

$$\text{Lift}(X \to Y) = \frac{\text{Support}(X,Y)}{\text{Support}(X) * \text{Support}(Y)} \tag{4.14}$$

The range of Lift values theoretically ranges from 0 to positive infinity. However, in practice, the interpretation of Lift values can be categorized as follows. A Lift value exceeding 1 signifies a positive association between items. A Lift value precisely at 1 indicates no association beyond what would be expected by chance. Conversely, a Lift value below 1 suggests a negative association.

4.1.5 Overfitting and underfitting

Usually, the desired goal is to get a model that is well generalized on the whole training set and not specific details of specific data points. Usually, when the model fails to generalize it overfits. Overfitting happens when the model achieves a high training accuracy yet performs poorly when encountering unseen data. Conversely, underfitting happens when the model performs poorly on the training and testing set. This implies that the model has failed to learn any pattern from the dataset. Poor data quality, improper feature selection, few training samples, an imbalanced dataset, and a bad selection of training parameters often cause model overfitting and underfitting. Model overfitting and underfitting can be handled by using techniques such as data balancing, proper feature selection, data augmentation, and cross-validation.

4.1.6 Model optimization

Optimization is adjusting training parameters (i.e., model coefficients) to minimize errors made when mapping the inputs to outputs by the machine learning model. Adjusting training parameters (i.e., tuning) is usually required to build a model that performs well and yields accurate predictions for a particular problem. During optimization, a model uses a parameterized mapping function (e.g., a weighted sum of inputs) to learn and generalize from training data to predict new data. The optimization algorithm usually minimizes the function's error and generates the optimal parameters by selecting values that cause the trained model to provide the best performance. The algorithm compares the results in every iteration by changing the parameters in each step until it reaches an optimum set of values. The selection and adjustment of parameters directly and significantly impact how the model performs.

In implementing machine learning models, various techniques are pivotal for optimizing algorithms. This section delves into four prevalent and traditional optimization methods to offer a concise understanding. The techniques explored include Exhaustive Search, Gradient Descent, Stochastic Gradient Descent, and Evolutionary Optimization Algorithms. Each method is detailed in the following subsections.

4.1.6.1 Exhaustive search

Exhaustive or brute-force search involves finding the most optimal parameters by examining whether each value is a good match. An excellent example of an exhaustive search is when someone forgets the combination of the digits (code) for a suitcase lock and tries out all possible combinations of digits to unlock it. The same approach is applied in model optimization, but the number of possible options (i.e., parameters' combinations) is typically very large. First, it generates a list of parameters and their corresponding values. Then, it trains and evaluates a model for each parameter combination, selecting the one with the best performance based on a predefined metric. Examples of machine-learning algorithms that can be optimized using exhaustive search are K-means clustering, Fuzzy c-mean clustering, and kNN classification algorithms.

4.1.6.2 Gradient descent

Gradient refers to the slope or incline of a surface. Thus, gradient descent means a descending slope to reach the lowest point in a particular space. The idea of the gradient descent method is to update the model parameters iteratively to minimize the objective function, whose parameters are optimized during training. With every update, this method guides the model in finding the target and gradually converges to the optimal value of the objective function. More precisely, it first initializes model parameters randomly with predefined values. Then, it computes the gradient of the loss function with respect to each parameter using training data and adjusts the parameters accordingly to converge toward the optimal values that minimize the loss gradually. When performing parameter optimization, the gradient descent optimization technique utilizes all data samples in a given dataset in every iteration. Thus, performing optimization with a large dataset in each iteration becomes computationally very expensive. Examples of machine-learning algorithms that can be optimized using gradient descent are logistic regression, linear regression, SVM, gradient boosting, and AdaBoosting.

4.1.6.3 Stochastic gradient descent

In contrast to gradient descent, which uses all data samples from the dataset in every iteration, Stochastic Gradient Descent (SGD) uses a few samples (or a batch) that are selected randomly in each iteration. A batch refers to the complete set of samples from a dataset utilized to compute the gradient in every iteration. Thus, in SGD, the learning algorithm normally finds out the gradient of the objective function for a batch in each iteration rather than the sum of the gradients of the objective function of all the samples. Since only a batch from the dataset is randomly selected for each iteration, the time taken by the algorithm to reach the optimal performance is usually significantly shorter compared to gradient descent methods. Some of the algorithms that are optimized by using SGD include logistic regression and SVM.

4.1.6.4 Evolutionary optimization algorithms

Evolutionary Optimization Algorithms (EOA) are population-based methods inspired by biological principles employed in solving machine-learning optimization problems. These algorithms draw inspiration from natural phenomena such as natural selection, species migration, bird swarms, human culture, and ant colonies. EOA starts by initializing a population of potential solutions, where each solution is represented by individuals possessing sets of parameters. They then evaluate the fitness of each individual based on an objective function, selecting individuals based on their fitness and generating new candidate solutions through recombination and mutation operations. Offsprings are introduced to the population, either replacing or supplementing existing individuals. The process continues for multiple iterations until termination criteria are met, allowing individuals to evolve toward better solutions efficiently. Examples of EOA include Genetic Algorithms (GA), Ant Colony Optimization, and Particle Swarm Optimization. It is worth noting that while EOAs can optimize machine learning models effectively, they do not necessarily find the optimal solutions.

4.2 MODEL DEPLOYMENT

The deployment of the machine-learning model involves putting a trained and validated model into a working environment. The machine-learning models can be deployed across a wide range of environments, such as web and mobile platforms, and are often integrated with other systems through Application Programming Interfaces (API) to facilitate accessibility for end users. The process of deploying the model requires several different key steps. Firstly, the model needs to be deployed into its working environment, where it has access to the hardware resources and data to work on. Secondly, the model is made accessible to end users' devices. Finally, the end users are trained to interact with the model via a simplified interface where they can insert their inputs and receive corresponding outputs.

4.3 MODEL MONITORING

The deployed model is continuously monitored to ensure that it performs predictions properly. Apart from performance monitoring, it is also important to ensure that the API and computation resources perform as required. Additionally, the model's performance should be routinely assessed using tools that track metrics to automatically give alerts should there be any degradation in its performance. Common causes of performance degradation include:

- **Variance in Input Data**: The data given to the model might not be cleaned in the same way as it was for the training and testing data which could adversely affect the performance of the model.
- **Changes in Data Integrity**: Over time, changes in data (e.g., formats and attribute naming) being fed to the model can affect the model's performance.
- **Data Drift**: Changes in features like demographics and market shifts can lead to data drift. This makes the data used during training become irrelevant with respect to the current context thereby making the model's results less precise.

- **Concept Drift**: End users' perceptions of correct predictions may change over time, making the model's predictions less relevant.

4.4 ETHICAL CONSIDERATIONS IN MACHINE LEARNING OPERATIONS (MLOps)

Ethical considerations within MLOps entail a spectrum of principles and practices focused on ensuring fairness, transparency, privacy, accountability, security, and diversity in the development, deployment, and use of AI systems. These considerations are crucial for mitigating potential harms, preventing discrimination and bias, protecting individual rights and privacy, and promoting trust and accountability in AI technologies. Ethical frameworks and guidelines offer direction on navigating complex ethical challenges and ensuring the responsible and ethical development and deployment of AI systems. Table 4.3 summarizes common ethical considerations in MLOps.

TABLE 4.3 Common ethical considerations in MLOps.

ETHICAL CONSIDERATION	DESCRIPTION
Fairness and Bias	Ensure that algorithms avoid discriminating against individuals based on protected features such as religion, race, or gender.
Accountability and Responsibility	Holding developers and organizations accountable for the actions and outcomes of deployed models, including resolving any errors or biases that arise.
Diversity and Inclusion	Ensure varied representation in model development to avoid bias perpetuation and to create inclusive solutions for all individuals.
Privacy and Security	Implementing strong privacy and security controls to prevent unwanted access, alteration, or exploitation of deployed models and the data they handle.
Explainability and Interpretability	Creating interpretable and explainable algorithms is crucial to upholding accountability, nurturing trust, and uncovering potential biases. When users comprehend how an algorithm makes its decisions, it promotes transparency and trust.
Human-in-the-Loop Approaches	Implement a human-in-the-loop approach where human judgment is involved in critical decisions made by deployed models. Establish redress mechanisms for individuals who perceive algorithmic decisions have negatively impacted them.
Legal and Regulatory Compliance	Ensure that algorithms comply with pertinent legal and regulatory frameworks related to fairness and non-discrimination to prevent potential legal and ethical conflicts.
Continuous Evaluation and Improvement	Emphasize continuous evaluation of algorithms post-deployment. Regularly update models, reevaluate fairness metrics and incorporate improvements to address emerging challenges and issues.

4.5 SUMMARY

This chapter provided a comprehensive guide on developing, deploying, and monitoring machine learning models. It began by discussing the critical considerations in dataset splitting techniques, emphasizing the importance of partitioning data into training and testing sets to ensure effective model evaluation. Additionally, it covered strategies for choosing the appropriate algorithm based on factors like the nature of the problem, dataset characteristics, and available computing resources. After that, the chapter discussed the model training and evaluation steps, explaining how to build the model and emphasizing the importance of assessing model performance on unseen data to gauge generalization capabilities accurately. Subsequently, the chapter presented several evaluation metrics, including R-squared, Accuracy, Silhouette score, Support, etc., used to facilitate performance evaluation, ensuring consistency and reliability in real-world applications. Additionally, the chapter explored the concepts of overfitting and underfitting, along with their corresponding mitigation techniques. Furthermore, key algorithms such as Gradient descent and EOA were presented to provide a comprehensive understanding of model optimization. Then, the chapter discussed model deployment and monitoring, underscoring the significance of deploying models in production environments and continuously monitoring their performance to address potential drift and maintain efficacy in dynamic settings. Lastly, the chapter concluded by presenting ethical issues in MLOps, which encompasses various principles and practices that promote fairness, transparency, privacy, accountability, security, and inclusion in the development, deployment, and use of AI systems.

Exercises

1. While training a machine learning model, discuss the role of k-fold cross-validation in preventing overfitting.
2. Explain key considerations when selecting a machine learning algorithm for different problems.
3. Identify and explain scenarios where improper data splitting could result in a biased model.
4. Describe the primary steps involved in model training and testing in machine learning. Highlight the significance of testing sets for model performance evaluation.
5. Define overfitting and underfitting in the context of machine learning models. Discuss strategies to mitigate these issues in model development.
6. Examine the optimization techniques used in machine learning algorithms. Discuss how optimization impacts model performance and efficiency.
7. Explain the importance of model evaluation in machine learning. Describe commonly used evaluation metrics and their relevance in assessing model performance.

8. Describe the process of machine learning model deployment and monitoring. Highlight the key factors to consider when deploying a model into production and establishing monitoring systems.
9. Critically identify and discuss ethical considerations in machine learning operations.
10. What are the potential benefits and challenges of adopting MLOps practices within an organization, and how can these challenges be overcome?

FURTHER READING

Dangeti, P. (2017). *Statistics for machine learning.* Packt Publishing Ltd.
Gollapudi, S. (2016). *Practical machine learning.* Packt Publishing Ltd.
Hall, M. (2011). *Practical machine learning tools and techniques.* Morgan Kauffman.
Paleyes, Andrei, Urma, Raoul-Gabriel, & Lawrence, Neil D. (2022). Challenges in deploying machine learning: A survey of case studies. *ACM Computing Surveys* 55(6), 1–29.
Pruneski, James A., Williams, Riley J., Nwachukwu, Benedict U., Ramkumar, Prem N., Kiapour, Ata M., Kyle Martin, R., Karlsson, Jón, & Pareek, Ayoosh. (2022). The development and deployment of machine learning models. *Knee Surgery, Sports Traumatology, Arthroscopy* 30(12), 3917–3923.
Simon, D. (2013). *Evolutionary optimization algorithms.* John Wiley & Sons.
Singh, P. (2021). *Deploy machine learning models to production.* Springer.
Subasi, A. (2020). *Practical Machine Learning for Data Analysis Using Python.* Academic Press.
Thompson, S. (2023). *Managing machine learning projects: From design to deployment.* Simon and Schuster.
Witten, Ian H., & Frank, E. (2002). Data mining: practical machine learning tools and techniques with Java implementations. *Acm Sigmod Record* 31(1), 76–77.

Machine learning software and hardware requirements

Upon completing this chapter, learners should be able to:

1. Describe commonly used software tools and libraries in machine learning development, including TensorFlow, PyTorch, scikit-learn, and Apache Spark.
2. Evaluate different hardware options for machine learning tasks based on performance, cost, and scalability considerations.
3. Demonstrate proficiency in setting up and configuring machine learning environments, including software installation, package management, and virtual environments.
4. Understand the importance of software version control and collaboration tools (e.g., Git and GitHub) in machine learning projects.
5. Explore cloud-based machine learning platforms and services for scalable model training and deployment.

5.1 PROGRAMMING LANGUAGES

It is important to acknowledge that proficiency in computer programming is essential for developing machine learning models. Python, R, and MATLAB are widely recognized as prominent programming languages in this field. They offer comprehensive software tools, including frameworks, Integrated Development Environments (IDEs), and libraries designed to construct machine learning models. These languages boast large and active communities comprising developers, data scientists, researchers, and enthusiasts. These communities contribute to advancing libraries, providing assistance, and sharing knowledge and resources. They offer extensive documentation, tutorials, forums, and online courses that facilitate learning. Moreover, Python, R, and MATLAB are user-friendly, readable, and versatile, which makes them accessible to both beginners and experienced developers. This fosters a supportive and collaborative environment within these communities. The following subsections provide in-depth discussions of the programming languages commonly used in machine learning.

5.1.1 Python programming language

Python is a versatile, object-oriented, open-source programming language widely used for crafting machine learning models. Its flexibility allows the implementation of various machine learning models through a range of Python-based software tools. Unlike Windows, Linux and Mac operating systems come with a Python environment pre-installed by default. Presently, Python exists in two primary versions: Python 2.x and Python 3.x, where x represents a minor version within the primary versions. Thus, offering distinct functionalities and features allows users to choose between versions based on specific project requirements and compatibility needs. The latest Python version is 3.x and can be installed through a Python setup file (available at: https://www.python.org/downloads/). The goal of installing Python is to create an environment that supports Python code execution. Python can also be automatically installed when installing other software packages such as Anaconda (available at: https://www.anaconda.com/).

5.1.1.1 Python code editors and IDEs

Most Python tools come packaged in a single distribution platform called Anaconda. Anaconda is an open-source platform and environment manager with several open-source packages (i.e., libraries, IDEs, and editors), as shown in Figure 5.1. Additional libraries to

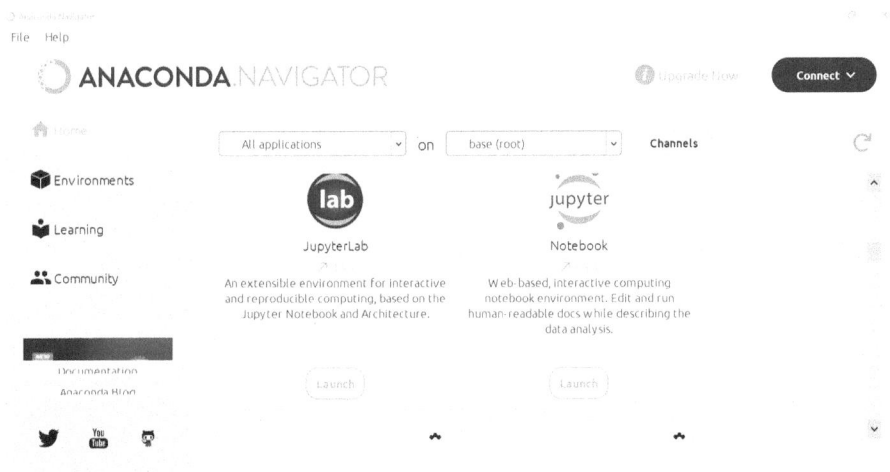

FIGURE 5.1 Anaconda environment.

Anaconda can be installed using Anaconda's package managers. Code editors and IDEs that support Python are needed to write the Python programs. A code editor is a text editor that simplifies and accelerates code writing and editing processes. On the other hand, an IDE is a software application used for creating, compiling, and debugging code. The common code editors and IDEs that support the Python programming language are summarized in Table 5.1.

TABLE 5.1 Python code editors and IDEs

TOOL	DESCRIPTION
Jupyter Notebook	Jupyter Notebook is a web-based open-source application offering an intuitive and interactive platform for data exploration, model development, visualization, documentation, and collaboration. Instead of composing and revising an entire program, Jupyter Notebook enables users to iterate and write Python code lines within cells, executing them individually. It facilitates easy modifications by allowing users to jump to cells, edit their code, and rerun the program seamlessly.
JupyterLab	JupyterLab presents the evolution of the Jupyter Notebook, offering an upgraded and more versatile interface for data exploration and computational tasks. It retains the core functionalities of the Notebook while introducing an enhanced user interface that allows for improved data analysis, visualization, and workflow organization.
PyCharm	PyCharm is an IDE that allows code completion and inspections, error highlighting and fixes, debugging, version control, and code refactoring. The major drawback of this software is that it is resource-intensive.
Spyder	Spyder is used for Python program development and has autocompletion, debugging, and variable exploration features. It has an area for writing Python code, a console, and a place for displaying variables, plots, and files.
Visual Studio Code (VS Code)	VS Code is a versatile code editor supporting numerous programming languages like Python, C++, PHP, and more. Its wide array of features and extensions makes it an excellent option for ML model development, testing, and deployment. These features encompass IntelliSense for intelligent code completion, integrated debugging tools, Jupyter Notebooks support, and an extensive library of extensions, streamlining Python development in machine learning endeavors.
Sublime Text	Sublime Text is a lightweight, cross-platform code editor known for its simplicity, speed, and user-friendly interface. It supports multiple programming and markup languages and offers many robust editing features, such as syntax highlighting, code folding, auto-completion, multiple selections, and macros. It is important to note that Sublime Text is not integrated into Anaconda.
PyDev	PyDev is a prominent open-source Python IDE built in the Eclipse platform. It has various capabilities, such as code completion, syntax highlighting, debugging tools, and integration with major Python libraries and frameworks.
Wing	Wing is a powerful proprietary IDE with open-source community editions. It includes advanced capabilities such as code analysis, debugging tools, integrated profiling, and support for various frameworks and libraries, including Python.

TABLE 5.1 (*Continued*)   Python code editors and IDEs

TOOL	DESCRIPTION
Geany	Geany is a code editor that is lightweight and efficient and can be used for Python programming. It has syntax highlighting, code folding, and project management features.
Brackets	Brackets is an open-source code editor intended mostly for web development but also suited for Python programming, including live preview, preprocessor support, and task-specific extensions.

5.1.1.2 Python libraries

Python offers a vast selection of libraries explicitly designed for constructing machine learning models. These libraries cover various functionalities, including classification, regression, clustering, collaborative filtering, dimensionality reduction, and optimization algorithms. Depending on the Python environment, these libraries may come pre-installed or can be readily accessed and installed using commands like *conda* and *pip* (i.e., Conda (*conda install <library name>*) or Pip (*pip install <package name>*)). A notable benefit is that most of these libraries are open-source, providing users with the flexibility to utilize and customize them at no cost. This significantly contributes to the collaborative and innovative environment for machine learning research and development. Table 5.2 outlines the most prevalent Python libraries.

5.1.2 R programming language

R is a no-cost, open-source programming language and environment devised for statistical computing and model creation. It boasts numerous capabilities, including robust techniques for data cleaning, transformation, integration, and preprocessing. Additionally, it offers various statistical tools, such as the chi-square test, t-test, z-test, and ANOVA, alongside machine learning tools like regression, classification, and clustering modeling. The R environment can be installed on Windows, Linux, and Mac operating systems via a standalone software package called R Studio. For example, in Windows operating systems, R Studio (Figure 5.2) can be installed using the setup file downloadable from https://cloud.r-project.org/ or through the Anaconda distribution platform.

5.1.2.1 R programming code editors and IDEs

Several popular IDEs and code editors support the R programming language, offering diverse options for users. Notable ones include Jupyter Notebook, Spyder, and VS Code, highlighted in Table 5.1. Each platform provides a robust environment for R programming, catering to different preferences and requirements, thereby accommodating a wide range of users and their varying workflow needs. Other IDEs and code editors that support the R programming language are presented in Table 5.3.

TABLE 5.2 Common Python libraries

LIBRARY	DESCRIPTION
NumPy	NumPy is a library employed for manipulating large, multidimensional arrays and matrices coupled with a suite of high-level mathematical functions tailored to operate on these arrays and matrices.
Pandas	Pandas are a powerful tool for loading, analyzing, and refining datasets, offering robust data manipulation and preparation functionalities. Leveraging the foundation provided by the NumPy library, Pandas extends its capabilities, providing a high-level interface and specialized tools for efficient data handling, transformation, and exploration.
Matplotlib	Matplotlib is a plotting library utilized for generating static, animated, and interactive 2D and 3D visualizations. It is commonly employed in conjunction with the NumPy library.
Seaborn	Seaborn is a Python data visualization library constructed atop Matplotlib. Offering a high-level interface enables the creation of visually engaging and informative statistical graphs, including scatter plots, line plots, histograms, box plots, and heatmaps. Seamlessly compatible with data frames and arrays, Seaborn aids in visually exploring and comprehending data.
scikit-learn	scikit-learn is employed for modeling tasks such as classification, regression, clustering, and dimensionality reduction. It incorporates a diverse range of machine learning algorithms, including SVM, Random Forest, and kNN, among others, for model development.
TensorFlow	TensorFlow is an end-to-end framework with a flexible ecosystem of tools, submodules, APIs, and community resources, aiding in developing and deploying classical machine learning and neural network–based models.
PyTorch	PyTorch is a library commonly used for developing and training neural network–based models. It is primarily developed to accelerate the path from prototyping to deployment.
Keras	Keras offers a high-level interface for creating and training deep learning models, enabling users to effortlessly design intricate neural networks with minimal coding, utilizing the capabilities of TensorFlow.
fastai	fastai is a deep learning library constructed atop PyTorch, aiming to streamline the training of deep learning models through user-friendly APIs and pre-trained models.
Plotly	Plotly is a library designed for crafting interactive and dynamic visualizations, providing a high-level interface to produce interactive plots, charts, and dashboards for tasks such as data exploration, model evaluation, and result presentation.
Plotnine	Plotnine is a Python library that applies graphics grammar to generate statistical graphs. Inspired by the ggplot2 package in R, it adopts a similar syntax and philosophy for visualization construction. With Plotnine, users can generate a diverse array of plots, including scatter plots, line plots, bar plots, histograms, and more, by mapping data attributes to aesthetic properties like color, shape, and size.
SciPy	SciPy library provides advanced scientific computing capabilities like optimization, integration, and linear algebra. These capabilities are essential for tasks such as signal processing and numerical analysis.
OpenCV	OpenCV is a computer vision library focused on image processing, feature detection, and object recognition.

TABLE 5.2 (*Continued*) Common Python libraries

LIBRARY	DESCRIPTION
Natural Language Toolkit (NLTK)	NLTK is a Python library dedicated to natural language processing tasks, providing extensive tools and resources for text processing, linguistic analysis, and machine learning.
Gensim	Gensim is a library focused on topic modeling and natural language processing (NLP), making it especially adept for text analysis tasks like document clustering and topic discovery.
Explain Like I'm 5 (ELI5)	ELI5 is a library for explaining machine learning models in simple terms, assisting in interpreting model predictions and gaining insights into their thinking.

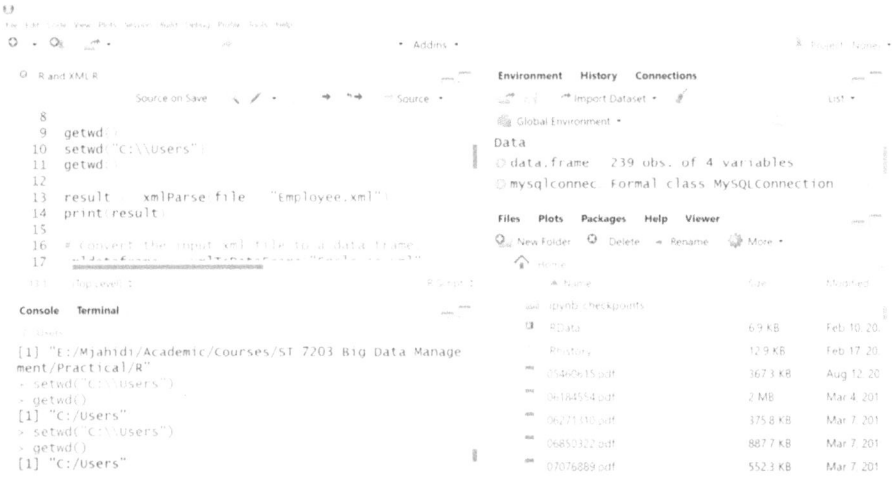

FIGURE 5.2 R studio environment.

5.1.2.2 *R programming libraries*

Several R libraries play crucial roles in building machine learning models, as detailed in Table 5.4. These libraries include different functionalities to address distinct aspects of machine learning tasks.

5.1.3 MATLAB

MATLAB® is a high-level programming language employed to express data or features in matrix and array form. It furnishes interactive tools, facilitating various machine learning tasks, including feature extraction, feature selection, model training, and hyperparameter tuning. As depicted in Figure 5.3, MATLAB offers diverse capabilities for managing machine learning tasks. It is worth noting that MATLAB is proprietary

TABLE 5.3 IDEs and code editors for R

TOOL	DESCRIPTION
RStudio	RStudio furnishes extensive tools and functionalities to aid R development, data analysis, and statistical modeling.
IntelliJ IDEA with R Plugin	IntelliJ IDEA is a Java IDE that supports R programming through its R plugin. It provides features such as code completion, debugging tools, and version control integration, offering a robust and reliable environment for R programming.
Eclipse with StatET	StatET plugin extends Eclipse's capabilities to support R programming. The plugin enhances Eclipse by incorporating syntax highlighting, code completion, and an integrated R console.
R Tools for Visual Studio	R Tools for Visual Studio is an extension of the Microsoft Visual Studio IDE enabling R programming. It provides various features, including IntelliSense, debugging, charting, remote execution, and SQL integration.
Atom	Atom provides a set of features, including syntax highlighting, code completion, debugging tools, an interactive console, data visualization capabilities, and project management functionalities.

TABLE 5.4 R programming libraries

LIBRARY	DESCRIPTION
DataExplorer	DataExplorer is a library used for EDA, feature engineering, and data reporting.
Ggplot2	Ggplot2 is a data visualization library renowned for producing visually appealing and informative plots, simplifying the exploration and communication of complex data patterns.
Kernel-Based Machine Learning Lab (kernLab)	kernLab is utilized for machine learning modeling tasks, encompassing classification, regression, clustering, and dimensionality reduction. It includes a diverse range of machine learning algorithms like SVM, Random Forest, and kNN.
MICE Package	Multivariate Imputation by Chained Equations (MICE) Package is used for imputing missing values in a dataset.
Rpart	Recursive partitioning (rpart) is a library used for classification, regression, and tree-based models.
Caret	caret offers a consolidated interface for training and assessing an extensive array of classification and regression models. The library streamlines the tasks of model selection, hyperparameter tuning, and performance evaluation.

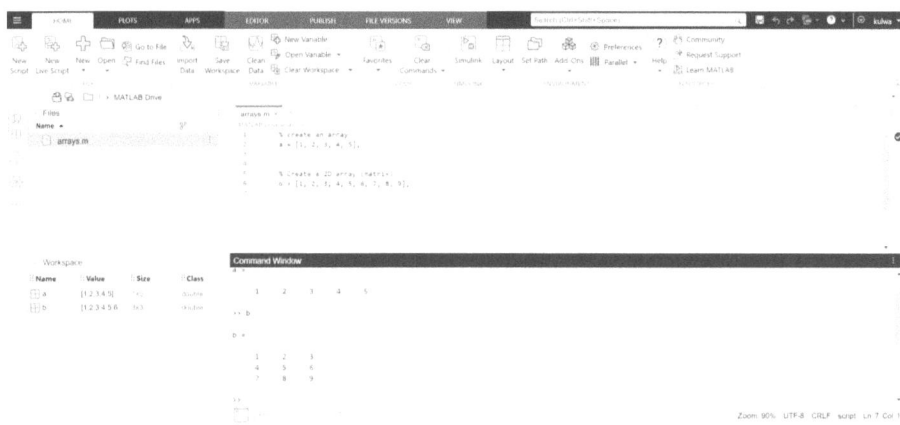

FIGURE 5.3 MATLAB working environment.

software compatible with Windows, Linux, and Mac operating systems. Further information on MATLAB installation can be found at https://www.mathworks.com/help/install/install-products.html.

5.1.3.1 MATLAB code editors and IDEs

MATLAB does not have a wide variety of code editors and IDEs, unlike Python and R. MATLAB Desktop is the primary and most widely used IDE for MATLAB programming. It has an interactive editor, command window, debugger, and various toolboxes for numerical computation, visualization, and programming. MATLAB Online and MATLAB Mobile are the web- and mobile-based versions of MATLAB Desktop, respectively, offering the same functionalities as the desktop version.

5.1.3.2 MATLAB libraries

Several MATLAB programming libraries are used for building machine learning models as summarized in Table 5.5.

5.1.4 Other programming languages

Python, R, and MATLAB are popular choices for machine learning, as described in the previous sections. However, other languages like Java and C++ can also be used as discussed in the following.

5.1.4.1 Java programming

Java is one of the predominant programming languages in the Information and Communication Technology (ICT) domain, renowned for its platform agnosticism, readability, and vast ecosystem. Machine learning in Java remains significant for various

TABLE 5.5 MATLAB programming libraries

LIBRARY	DESCRIPTION
MATLAB Image Processing Toolbox	The MATLAB Image Processing Toolbox offers comprehensive functions and tools for processing, analyzing, and visualizing images.
MATLAB Signal Processing Toolbox	The MATLAB Signal Processing Toolbox comprises functions tailored for signal analysis, filtering, feature extraction operations, and spectrum analysis.
MATLAB Statistics and Machine Learning Toolbox	The MATLAB Statistics and Machine Learning Toolbox provide functions and algorithms for statistical analysis, machine learning, and predictive modeling. It provides functionalities for classification, regression, clustering, and dimensionality reduction.
MATLAB Optimization Toolbox	MATLAB Optimization Toolbox contains a set of algorithms and tools for handling optimization issues such as linear programming, nonlinear optimization, and restricted optimization.
MATLAB Curve Fitting Toolbox	MATLAB Curve Fitting Toolbox includes tools for fitting curves, interpolating data, and smoothing data. It provides a variety of curve-fitting methods as well as tools for analyzing and displaying fitted curves.

reasons, including its wealth of libraries and frameworks, seamless integration with existing Java codebases, robust performance and scalability, applicability in enterprise environments, emphasis on security, versatility across diverse use cases, compatibility across multiple platforms, and strong community support.

5.1.5 Java programming code editors and IDEs

Java programming code editors and IDEs for machine learning development provide advanced syntax highlighting, code completion, and debugging capabilities specifically designed for Java machine learning libraries. These tools streamline the machine learning workflow by incorporating version control systems like Git and granting convenient access to libraries and frameworks for tasks such as data preprocessing, model training, and evaluation. Moreover, they boast a diverse ecosystem of plugins and extensions for additional customization, enhancing productivity in machine learning projects. These tools are outlined in Table 5.6.

5.1.6 Java ML libraries

Java has specialized machine learning libraries offering various functions, from data preparation to model evaluation. By leveraging these libraries, developers can better utilize Java's robust ecosystem to build and deploy machine learning solutions. Some of the libraries are described in Table 5.7.

TABLE 5.6 IDEs and code editor for Java

TOOL	DESCRIPTION
IntelliJ IDEA	IntelliJ IDEA is an IDE tailored for Java development, facilitating the creation of robust code across various platforms such as Windows, macOS, and Linux. It offers two editions: a no-cost community version and a paid ultimate edition.
Eclipse	Eclipse provides both a desktop version and a cloud version known as Eclipse Che. This IDE empowers developers to manage multiple workspaces concurrently, enhancing project organization and boosting productivity and efficiency.
NetBeans	NetBeans is a cross-platform, open-source IDE designed for Java development that is free of charge. Packed with features like syntax highlighting, code completion, and integrated debugging tools, the IDE facilitates rapid coding.
BlueJ	BlueJ, a free IDE commonly utilized for educational aims, is particularly beginner-friendly. This well-organized platform offers an interactive environment complemented by graphical representations and a distinctive coloring scheme.
JDeveloper	JDeveloper, a free IDE, is particularly suited for streamlining Java application development across the System Development Life Cycles (SDLC). This tool stands out for its features, including advanced code editing functionalities, seamless integration with version control systems like Git, automated deployment tools, and strong support for Java technologies like Enterprise JavaBeans (EJB) and Java Persistence API (JPA).
JCreator	JCreator is a versatile IDE suitable for developers of all levels (i.e., deals for beginners and experienced professionals), with a lightweight design and robust features. It offers an intuitive interface, advanced code editing, integrated debugging, project management tools, version control integration, GUI design, profiling, code analysis, and support for plugins.
Codenvy	Codenvy is a cloud-based software that allows developers to work and collaborate without installing local software on their machines. This makes it ideal for remote software development teams that need a unified platform that their global workforce can use to work individually and collaborate.
DrJava	DrJava is a lightweight, user-friendly IDE primarily designed to offer simplicity and ease of use for developers. It is particularly preferred in educational settings due to its beginner-friendly interface and features tailored for learning Java programming.
JGrasp	JGrasp, a straightforward Java IDE, is particularly commendable for educational purposes. It boasts syntax highlighting, code navigation, and UML visualization capabilities, all packaged within a user-friendly interface that facilitates the automatic creation of software visualizations. Notably, it specializes in generating Control Structure Diagrams (CSDs), technical diagrams crucial for illustrating control flow in applications. This functionality aids debugging and workbench testing phases by enhancing developers' code readability.

TABLE 5.7 Java ML libraries

LIBRARY	DESCRIPTION
TensorFlow Serving	TensorFlow Serving, an open-source library, is tailored for deploying machine learning models focusing on achieving low latency performance. It can operate locally or in cloud environments, accommodating a wide array of models, ranging from deep convolutional networks to linear models. This tool empowers developers to efficiently deploy machine learning models at scale, eliminating the need for manual infrastructure management.
Apache Spark MLlib	Apache Spark MLlib, a specialized library is crafted to construct machine learning pipelines within Apache Spark clusters. Equipped with high-level APIs, it empowers developers to swiftly establish resilient machine learning pipelines by leveraging distributed data training algorithms and other distributed processing tasks.
DL4J	Deeplearning4j (DL4J), a robust deep learning library, is constructed atop the Java Virtual Machine (JVM), aiding developers in crafting production-ready applications. With provisions for GPU acceleration, distributed computing, and diverse neural network architectures like convolutional nets, recurrent neural nets, and LSTM networks, DL4J ensures comprehensive support. Additionally, it offers a GUI-based user interface for hyperparameter tuning, simplifying the optimization of model performance.
Apache OpenNLP	The Apache OpenNLP library specializes in natural language processing (NLP) tasks within the Java environment. With functionalities like tokenization, part-of-speech tagging, sentence segmentation, and named entity recognition, OpenNLP offers a modular architecture and pre-trained models, streamlining the integration of NLP features into Java applications.
Apache Mahout	Apache Mahout is a Java library tailored to deliver scalable machine learning algorithms, covering clustering, classification, and recommendation tasks. Engineered to handle extensive datasets efficiently, Mahout excels in performing machine learning operations on big data.
Smile	The Statistical Machine Intelligence and Learning Engine (Smile) is a Java library featuring various algorithms for classification, regression, clustering, association rule mining, and dimensionality reduction. With a focus on simplicity and performance, Smile offers an intuitive API suitable for novice and experienced developers.
TensorFlow Java API	TensorFlow, a renowned deep learning library, offers a Java API that enables developers to integrate TensorFlow capabilities seamlessly into Java applications. This facilitates the development and training of neural networks within Java environments.

TABLE 5.7 (*Continued*) Java ML libraries

LIBRARY	DESCRIPTION
DL4J	Deep Learning for Java (DL4J) is a distributed deep learning library designed for Java, Scala, and Clojure. It harmonizes with Hadoop and Spark, accommodating diverse neural network architectures.
Encog	Encog emerges as a sophisticated machine learning framework tailored for Java, encompassing neural networks, genetic algorithms, support vector machines, and various other ML techniques.
JSAT	Java Statistical Analysis Tool (JSAT) is a Java-based library housing ML algorithm, prioritizing user-friendliness and mirroring the design of the Weka library. JSAT offers an extensive array of algorithms for classification, regression, clustering, and recommendation, suitable for researchers, students, and enthusiasts keen on experimenting with ML algorithms in Java.
MALLET	MALLET, which stands for Machine Learning for Language Toolkit, is a Java-based toolkit designed for natural language processing (NLP), encompassing tasks such as document classification, clustering, topic modeling, and information extraction. Renowned for its flexibility, user-friendly interface, and comprehensive documentation, MALLET is accessible to novices and seasoned NLP practitioners.

5.1.6.1 C++ programming

C++ is a versatile and powerful programming language widely utilized in machine learning to develop core algorithms and implement computationally intensive tasks. With its high speed, efficiency, reliability, and low-level control, C++ caters to diverse domains beyond machine learning, including game development, embedded systems, and software engineering. This ability is attributed to its support for procedural and object-oriented programming paradigms and low-level memory manipulation features.

5.1.7 C++ programming code editors and IDEs

C++ code editors and IDEs are tools that offer a variety of features tailored for C++ development. These tools support code editing, debugging capability, and integration with C++ libraries essential for ML-based project management. The C++ code editors and IDEs are described in Table 5.8.

5.1.8 C++ programming libraries

C++ offers numerous libraries tailored for machine learning and AI applications, equipped with pre-implemented algorithms, functions, and tools to construct intelligent systems. Table 5.9 outlines some of the prominent libraries for machine learning in C++.

TABLE 5.8 C++ programming code editors and IDEs

TOOL	DESCRIPTION
Code::Blocks	Code::Blocks is a free, cross-platform IDE tailored for C/C++ development, offering a range of features like compiling, debugging, profiling, and code analysis. Renowned for its performance and user-friendly interface, it supports full breakpoints and integrates seamlessly with community and team-developed plugins.
CodeLite	CodeLite is also an open-source IDE that comes with the features of a class browser, static code analysis, project management, code refactoring, profiling, debugging, completion, and compiling. The IDE offers a rapid application development (RAD) tool that helps one build widget-based applications. Windows, Linux, Mac, and FreeBSD support it.
CLion	CLion is a cross-platform IDE built for C++ development, providing features such as code analysis, CMake support for streamlined project management and build automation, and intelligent code assistance for project modeling. Notably, it offers local and remote (via SSH) support, enabling developers to code locally and compile on remote servers.

TABLE 5.9 C++ programming libraries

LIBRARY	DESCRIPTION
Dlib	Dlib, an open-source, cross-platform toolkit, is primarily employed for machine learning and computer vision applications. Renowned for its high performance and efficiency, it provides many tools and algorithms for facial recognition, object detection, image processing, and machine learning model training, making it ideal for real-time applications.
mlpack	mlpack is a versatile machine learning library designed to provide state-of-the-art algorithms for clustering, regression, and dimensionality reduction, along with data preprocessing and visualization tools. Utilizing the Armadillo linear algebra library, mlpack emphasizes scalability, speed, and user-friendliness, making machine learning model development accessible to novice users through a simple and consistent API.
SHARK	SHARK is a collection of open-source C++ machine learning libraries that offer linear and nonlinear optimization, kernel-based learning algorithms, neural networks, and various machine learning methods. It empowers machine learning experts to easily tackle a broad spectrum of tasks, making it suitable for real-world applications and research endeavors. SHARK's versatility extends to supervised and unsupervised learning, evolutionary algorithms, and other machine learning techniques, providing a robust toolkit for diverse machine learning challenges.
Caffe	Caffe, developed by the Berkeley Vision and Learning Center (BVLC), is a high-performance deep learning framework designed for the efficient training and deployment of neural networks, particularly in computer vision. Its modular architecture facilitates experimentation, and its CPU and GPU acceleration support allows it to handle large-scale machine learning tasks efficiently. Caffe is rich in pre-trained models and visualization tools, making it popular among deep learning researchers and practitioners.

TABLE 5.9 (Continued) C++ programming libraries

LIBRARY	DESCRIPTION
CNTK	The Microsoft Cognitive Toolkit (CNTK) is an open-source platform for distributed deep learning, known for its high accuracy in training deep learning models. It features a flexible and powerful API for C++.
Armadillo	Armadillo is a robust C++ linear algebra library with MATLAB-like syntax and functionality, simplifying matrix, linear algebra, and numerical tasks. Its intuitive interface enhances development productivity, while seamless integration with other C++ libraries makes it versatile for scientific computing, machine learning, and data analysis. Known for its speed, ease of use, and compatibility, Armadillo is favored in academic research and industrial applications requiring fast and reliable numerical computations.
DyNet	DyNet is a C++ library with Python bindings optimized for dynamic computation graphs and automatic differentiation. It excels in neural network operations and training, particularly in natural language processing tasks where it is frequently applied.
Shogun	Shogun offers various machine learning algorithms and tools for classification, regression, clustering, and dimensionality reduction tasks. With bindings for Python, Java, and MATLAB, users can access its functionalities from various programming environments despite its core implementation being in C++.
FANN	Fast Artificial Neural Network (FANN) is an open-source neural network library written in C language (it also supports C++). The library implements multilayer artificial neural networks supporting fully and sparsely connected networks. It is easy to use, versatile, well-documented, and fast. Critical features of FANN include backpropagation learning, evolving topology learning, cross-platform, and support for floating and fixed point numbers.
FAISS	FAISS offers efficient algorithms for similarity search and clustering of dense vectors. With Python bindings, it integrates well with Python-based machine learning workflows. Its core functions are in C++, ensuring high efficiency for tasks like large-scale nearest-neighbor search. FAISS supports CPU and GPU acceleration, making it versatile for applications like image and text retrieval, recommendation systems, and NLP.
OpenNN	OpenNN supports machine learning and advanced analytics across various domains like energy, marketing, health, and digital economy. With algorithms for classification, regression, and prediction, OpenNN offers robust AI solutions. Its multiprocessor programming ensures high performance for the swift execution of complex tasks.

5.1.9 Criteria for choosing programming language for machine learning

When choosing a programming language for machine learning projects, key factors include library and framework support, robust and extensive community support, ease of learning and use, flexibility, scalability and efficiency, integration with other tools and software, and industry adoption. Languages like Python are favored for their simplicity and extensive ecosystem of machine learning libraries, while languages like C++ and Java excel in performance-intensive tasks. Python's interoperability and lightweight deployment options make it popular for integrating machine learning models into production systems. Ultimately, the choice depends on project requirements and development team preferences, with careful consideration of these factors ensuring the most suitable language is selected for machine learning projects. These criteria are discussed in the subsequent sections.

5.1.9.1 Library and framework support

Libraries are compilations of pre-written code modules that developers can utilize to save time and avoid reinventing the wheel. In AI and machine learning, where specific functionalities can significantly speed up the development process, libraries play a crucial role by offering ready-to-use algorithms and data structures. A programming language equipped with a diverse and robust set of libraries is often favored for AI and machine learning development. On the other hand, a framework is a pre-established, reusable toolkit comprising tools, libraries, and conventions. It serves as an abstraction layer, streamlining the development and maintenance of software applications by providing common functionalities, design patterns, and components. Robust library and framework support in a programming language can simplify and accelerate the execution of machine learning projects.

5.1.9.2 Robust and extensive community support

The presence of robust and extensive community support is crucial for navigating the challenges encountered while developing machine learning applications. Additionally, a large, active, and knowledgeable community associated with a particular programming language plays a pivotal role in selecting the language for machine learning projects. Such a community actively engages in discussions, forums, and online platforms, readily sharing expertise and knowledge. Moreover, it facilitates in-person connections through meetups and events, fostering experience exchanges among the members. A vibrant community benefits developers of all levels, enabling continuous learning and exposure to best practices. In the context of machine learning projects, programming language community support ensures resilience and sustainability by offering the members reliable assistance and shared knowledge. The active participation and extensive support from the community ultimately contribute to the success of machine learning endeavors.

5.1.9.3 Ease of learning and use

The ease of learning and use depends on factors such as user experience, familiarization with the programming language, or its direct impact on solving the problem. A programming language with high ease of learning has clear and concise documentation, a simple and consistent syntax, and features that make common tasks straightforward. Additionally, the availability of learning resources, community support, and a supportive development environment contribute to the overall ease of use. A programming language designed for ease of learning and use can accelerate the development process and reduce the likelihood of errors, making it more accessible and appealing to both beginners and experienced developers.

5.1.9.4 Flexibility, scalability, and efficiency

Choosing the right programming language for machine learning projects involves assessing flexibility, scalability, and efficiency to meet diverse needs and challenges throughout the development life cycle. A flexible language allows developers to write adaptable code that addresses various requirements by supporting multiple programming paradigms, offering diverse libraries and frameworks, and enabling concise expression of ideas. Scalability is crucial for accommodating growth in users, data, and features, requiring vertical and horizontal scaling capabilities. Support for parallel processing, efficient memory management, and distributed computing enhances a language's scalability. Additionally, efficiency is essential for executing tasks quickly and utilizing system resources effectively. Considerations such as runtime performance, memory management, and optimization tools are crucial for resource-intensive machine learning applications. Choosing a programming language that balances flexibility, scalability, and efficiency ensures robustness and adaptability in machine learning development.

5.1.9.5 Integration with other tools and software

Effective integration with other tools and software is crucial in selecting a programming language for machine learning projects. Seamless integration streamlines workflows, leveraging existing tools and infrastructure efficiently. Key considerations include robust APIs and libraries, compatibility with existing tools and frameworks, support for standard data exchange formats, efficient interprocess communication mechanisms, database integration, deployment, and cloud services. By considering these factors, developers can choose a programming language that facilitates seamless integration, enhances productivity, and maximizes efficiency in machine learning projects.

5.1.9.6 Industry adoption

Industry adoption is a pivotal factor influencing the choice of programming languages for ML projects. The widespread adoption of a language across various sectors signifies its relevance and suitability for real-world applications. One of the primary advantages of selecting a language with high industry adoption is the market demand it generates. Such languages are often in high demand, increasing job opportunities and career prospects for proficient developers. Moreover, industry adoption ensures the availability of a skilled talent pool. Companies prefer languages with a large community of proficient developers, making recruiting and onboarding talent with relevant expertise easier. Additionally, languages with extensive industry adoption typically enjoy stable ecosystems with robust support from developers, communities, and organizations. This stability ensures continuous development, updates, and maintenance support, reducing the risk of project disruptions.

5.2 NO-CODE TOOLS

No-code tools come with pre-packed implementations for common machine learning algorithms for classification, clustering, regression, dimensionality reduction, etc. They are used to quickly build machine learning models without requiring prior programming knowledge and skills. However, using programming languages to develop machine learning models is a better option than no-code tools because the former provides control over the created model. The common no-code tools are WEKA, Orange, and Teachable Machine. Table 5.10 summarizes common no-code tools for building machine learning models.

5.3 EXPERIMENT TRACKING TOOLS

Experiment tracking within machine learning encompasses the comprehensive management of all experiment components, from hyperparameters and performance metrics to predictions, ensuring the creation of an efficient and well-documented model. Table 5.11 presents a compilation of the commonly utilized tools designed explicitly for experiment tracking in machine learning. These tools offer diverse functionalities that aid in organizing, monitoring, and evaluating various experiment elements, contributing significantly to the streamlined development and optimization of machine learning models.

5.4 PRE-TRAINED MODELS REPOSITORIES

A pre-trained model is a solution developed for a specific problem, which can be directly applied or fine-tuned to address similar tasks. Leveraging pre-trained models can reduce computing costs, reduce carbon footprint, and save time on training models

TABLE 5.10 Common no-code tools

TOOL	DESCRIPTION
WEKA	WEKA, short for Waikato Environment for Knowledge Analysis, is a free and open-source tool for machine learning. It offers a range of algorithms for tasks like data preprocessing, classification, regression, clustering, association rules, and visualization. Further information on installing WEKA can be found at https://waikato.github.io/weka-wiki/downloading_weka/.
Orange	Orange is a free open-source toolkit for data visualization and machine learning featuring comprehensive libraries. It is conveniently included in the Anaconda distribution.
Teachable Machine	Teachable Machine is a free web-based machine learning no-code tool for easily prototyping models.

TABLE 5.11 Experiment tracking tools

TOOL NAME	DESCRIPTION
Dashboard by weight and biases	Weight and Biases Dashboard allows for the real-time monitoring of training data. It seamlessly integrates with popular machine learning frameworks like PyTorch, TensorFlow, and Keras.
Tensorboard	Tensorboard enables the visualization of statistics of a neural network, such as the training parameters (e.g., loss, accuracy, and weights), images, and even the graph to debug and optimize the model.
Neptune.ai	Neptune.ai is a centralized metadata repository for machine learning operations workflow, enabling tracking, visualization, and comparison of thousands of machine learning models in one place. It fosters seamless collaboration within the machine learning community.
MLflow	Machine Learning Flow (MLflow) is an open-source platform for managing the end-to-end machine learning life cycle. It has components for recording and querying experiments, packaging code into reproducible runs, managing and deploying machine learning models, supporting integration with popular machine learning frameworks and libraries, and storing and sharing machine learning models.
Comet ML	Comet Machine Learning (Comet ML) is a machine learning experimentation and collaboration platform. It can track, compare, and analyze experiments, log hyperparameters, metrics, and experiment results, and facilitate collaboration among team members. Comet ML also supports integration with popular machine learning frameworks and libraries.
Metaflow	Metaflow is a machine learning infrastructure tool developed by Netflix for building, deploying, and managing real-life data science projects by providing a high-level abstraction. It supports versioning, monitoring, and scaling machine learning pipelines. It enables users to define machine learning workflows as a series of steps and execute them locally or in the cloud.
ClearML	Clear Machine Learning (ClearML), previously called Trains, is an open-source platform designed to oversee machine learning experiments and models. It offers features for logging hyperparameters, metrics, and artifacts and tracking, visualizing, and optimizing machine learning workflows. Additionally, ClearML supports model deployment and monitoring, and seamless integration with popular machine learning frameworks and libraries.

TABLE 5.12 Pre-trained models' repositories

REPOSITORY NAME	DESCRIPTION
TensorFlow Hub	TensorFlow Hub contains pre-trained models that are available for deployment and fine-tuning. It facilitates the reuse of pre-trained models with a minimum amount of code added. This repository can be accessed at https://www.tensorflow.org/hub.
Pytorch Hub	PyTorch Hub provides a platform for publishing pre-trained models to a GitHub repository, including model definitions and pre-trained weights. This repository can be accessed at https://pytorch.org/hub/.
Hugging Face Transformers	The Hugging Face Transformers platform offers APIs that simplify downloading and retraining state-of-the-art pre-trained models. This repository can be accessed at https://huggingface.co/docs/transformers/index.
OpenAI	OpenAI provides powerful machine learning models created by OpenAI that are trained on massive quantities of data to reach outstanding language interpretation and generation skills. The pre-trained models' documentation can be accessed at https://platform.openai.com/docs/models.
Paperswithcode	Paperswithcode is a repository of machine learning research papers with links to the corresponding code and pre-trained models. It is helpful for researching cutting-edge models and locating suitable resources for specific requirements. It can be accessed at https://paperswithcode.com/.
OpenAI Model Zoo	OpenAI Model Zoo is a repository that contains a collection of high-performing pre-trained OpenAI models, including the GPT-3 family of big language models. It is well-known for its cutting-edge models. It can be accessed at https://platform.openai.com/docs/models.

from scratch. These machine learning models are readily available from established repositories, some of which are detailed in Table 5.12.

5.5 DATASETS AND MODEL TRACKING TOOLS

Datasets and model tracking tools are crucial in monitoring alterations made to datasets and gauging their influence on the performance of machine learning models. These tools are integral for tracking changes within the data and training and refining machine learning models. Table 5.13 provides a compilation of common datasets and model-tracking tools. Each tool within this compilation is pivotal in cataloging, managing, and analyzing datasets and monitoring the evolution of machine learning models, thereby aiding researchers and practitioners in effectively managing the intricate process of data modification and model refinement within the machine learning workflow.

TABLE 5.13 Datasets and model tracking tools

TOOL	DESCRIPTION
Artifacts by Weights and Biases	Artifacts by Weights and Biases are used to version the datasets, track different machine learning pipelines, and reproduce previous datasets.
Data Version Control (DVC)	DVC, an open-source version control system, is tailored to monitor and manage models and datasets within the machine learning workflow. It offers a structured framework for tracking changes in models and datasets, bolstering reproducibility and fostering collaboration among teams engaged in machine learning projects.
CML	Continuous Machine Learning (CML) is a GitHub Actions feature that allows one to automate machine learning activities, such as tracking datasets and model versions.
DataRobot MLOps	DataRobot MLOps platform offers the functionality for managing and tracking datasets, apart from the end-to-end machine learning life cycle.

5.6 AUTOML HYPERPARAMETER OPTIMIZATION TOOLS

Automated Machine Learning (AutoML) tools simplify the process of optimizing machine learning models by automatically adjusting their hyperparameters. Table 5.14 presents a compilation of the most commonly used tools for this purpose, providing a comprehensive overview of the techniques employed in AutoML hyperparameter optimization.

5.7 MACHINE LEARNING LIFE CYCLE TOOLS

Machine learning life cycle tools track every model development, deployment, and performance monitoring stage. They are used from the initial conception of the algorithm to the optimization, which is required to keep the model accurate and effective. The common machine learning life cycle tools are summarized in Table 5.15.

5.8 USER INTERFACE DEVELOPMENT TOOLS

The user interface is crucial in interactive machine learning as users actively train the algorithm iteratively. Table 5.16 compiles commonly used tools for developing interactive and effective interfaces.

TABLE 5.14 AutoML hyperparameter optimization tools

TOOL	DESCRIPTION
Optuna	Optuna is a freely available open-source framework developed explicitly for automatic hyperparameter optimization. Its user-friendly define-by-run API sets it apart, making the process more intuitive and adaptable to varying optimization needs.
Tune	Tune is a versatile Python library designed for experiment execution and automatic hyperparameter tuning, suitable for small- and large-scale machine learning projects. It facilitates efficient experimentation and parameter tuning across various task complexities.
HyperOpt	HyperOpt is a Python library for hyperparameter tuning that automatically chooses the best parameters for a given model. It is capable of optimizing large-scale models with hundreds of hyperparameters.
TPOT	TPOT, which stands for Tree-based Pipeline Optimization Tool, is a Python-based automated machine learning tool. It utilizes genetic programming to optimize machine learning pipelines automatically.
Google Cloud AutoML	Google Cloud AutoML is a tool developed by Google that automatically tunes hyperparameters in complex machine learning models.
AWS Sage Maker	AWS Sage Maker provides automatic optimization service to machine learning algorithms built using huge datasets in a distributed environment.
Microsoft (MS) Azure AutoML	MS Azure AutoML is a Microsoft-developed open-source toolkit for AutoML. It automates hyperparameter tuning, feature engineering, and model compression tasks.
Scikit-Optimize	Scikit-Optimize is an easy-to-use Python built-in library integrated with scikit-learn and provides basic hyperparameter optimization (HPO) algorithms such as grid search and random search.
Auto-PyTorch	Auto-PyTorch is a PyTorch models automation library focused on hyperparameter optimization (HPO), neural architecture search (NAS), and model pruning.
Auto-Keras	Auto-Keras is a specialized library integrated with Keras that focuses on automating neural architecture search (NAS) and hyperparameter optimization (HPO) specifically for Keras models.
IBM Watson AutoAI	IBM Watson AutoAI is a component of IBM Watson Studio that automates the training and optimization of ML models, including hyperparameter tuning.

TABLE 5.15 Machine learning life cycle tools

TOOL	DESCRIPTION
Kubeflow	Kubeflow is a free and open-source machine learning platform that facilitates the development, orchestration, optimization, deployment, and execution of scalable and portable models. It provides a framework for organizing projects, harnessing the power of cloud computing, and empowering developers to construct optimal models.
Seldom	Seldom is an open-source machine learning deployment platform that streamlines the machine learning workflow with features such as audit trails, advanced experiments, continuous integration, scaling, and model explanations, enabling faster and more effective problem-solving.
Mlflow	Mlflow is an open-source platform to manage the machine learning life cycle, including implementation, experimentation, packaging, deployment, and performance monitoring.
Google Cloud AI Platform	Google Cloud AI Platform provides a range of features for managing the machine learning life cycle. This includes a dashboard, data labeling, workflow orchestration, and model management.

TABLE 5.16 User interface development tools

TOOL	DESCRIPTION
Streamlit	Streamlit supports the development of web applications for machine learning problems. It is an open-source library's API written entirely in Python. Therefore, it simplifies web application development without utilizing other web technology languages.
Django	Django is a free and open-source framework for constructing web apps (i.e., user interfaces) based on Python programming. It is suitable for building secure, maintainable, and multi-page applications.
Flask	Flask is a Python-based microframework that offers basic features for developing web applications. It is suitable for single-page applications only.

5.9 EXPLAINABLE AI TOOLS

Explainable AI (XAI) tools provide detailed insights into the functioning of machine learning models through descriptive explanations. Table 5.17 is a convenient reference, showcasing these tools for easy understanding and practical application.

TABLE 5.17 Common XAI tools

TOOL	DESCRIPTIONS
SHAP	SHAP, which stands for SHapley Additive exPlanations, is a framework that offers explainability for various algorithms, including linear regression, logistic regression, and tree-based models. It provides insights into the contributions of individual features toward the predictions made by these models.
LIME	LIME, which stands for Local Interpretable Model-agnostic Explanations, is a methodology that provides explainability for a wide range of algorithms, including random forest, k-Nearest Neighbor (kNN), and support vector machines (SVMs). It enables the interpretation of individual predictions made by these models, allowing for a better understanding of their decision-making process.
AI Explainability 360	AI Explainability 360 is an open-source toolkit developed by IBM. It offers a comprehensive collection of techniques and models specifically designed for interpreting and explaining machine learning models. This toolkit provides a valuable resource for enhancing the transparency and interpretability of machine learning models.
Anchors	Anchors is a tool with simple, high-precision rules that locally characterize a model's behavior. These rules are interpretable and aid in comprehending the model's decisions.

5.10 VERSION CONTROL SYSTEMS

Version control systems (VCS) are software tools that track and manage file changes, enabling developers to record modifications in the source code. VCS maintains a repository of all changes, allowing developers to revert to earlier versions if needed. This facilitates error fixing and comparison of file versions. Moreover, VCS enables collaborative work by allowing multiple developers to edit files independently and share changes when ready. Table 5.18 provides examples of common VCS tools.

5.11 MACHINE LEARNING HARDWARE REQUIREMENTS

This chapter introduces the hardware requirements for machine learning tasks. It also introduces using cloud computing services as an alternative method if local computer resources do not meet the requirements of the machine learning process.

5 • Machine learning software and hardware requirements

TABLE 5.18 Version control systems

VERSION CONTROL SYSTEM	DESCRIPTION
Git	Git is an open-source distributed VCS designed to support projects of different sizes and support multiple branches of change that can be independent of each other.
Concurrent Versions System (CVS)	CVS is a free, open-source version control system that efficiently manages concurrent software development branches. It enables collaboration, tracks changes, and maintains version history.
Subversion (SVN)	SVN, an open-source version control system, offers a wide selection of Integrated Development Environment (IDE) plugins. These plugins enhance usability and integration with various IDEs, making version control seamless within the development environment. They facilitate smoother collaboration and code management among team members.
Mercurial	Mercurial is a distributed VCS with features similar to those of Git.
Data Version Control (DVC)	DVC, primarily designed for version management of machine learning projects, specializes in handling machine learning datasets and models. It efficiently handles large files, such as datasets, while effectively tracking changes.

5.12 OPERATING SYSTEMS REQUIREMENTS

The most commonly used operating systems in contemporary machine learning tasks include GNU/Linux-based OSs, Microsoft Windows, and Apple MacOS. However, modern machine learning algorithms primarily execute their computational tasks within the core software governing the entire computer system. Consequently, there is no inherent advantage in using a particular OS over others for the machine learning process. Moreover, considering additional factors such as the ease of supporting emerging technologies and the extensive support of free and open-source libraries, the Linux operating system holds more advantages than Microsoft Windows and MacOS.

5.13 PROCESSOR AND MEMORY REQUIREMENTS

Machine learning tasks often necessitate substantial computational resources due to the large datasets and complex algorithms involved. Selecting the most suitable machine for such tasks can be challenging, as several factors must be considered, including processing speed and graphics processing capabilities. The following subsections outline the minimum requirements for the Central Processing Unit (CPU), Graphics Processing Unit (GPU), Random Access Memory (RAM), and Storage to ensure optimal performance in machine learning workloads.

5.13.1 CPU

Multi-core processing, which involves distributing computationally intensive tasks across multiple CPU cores, is commonly employed in machine learning. Utilizing multiple cores can significantly reduce execution time, scaling performance gains with the number of available cores. A minimum recommendation for simple machine learning tasks would be a dual-core 2.2 GHz processor.

5.13.2 GPU

A GPU, or Graphics Processing Unit, is a specialized microprocessing chip or circuit for graphics-related tasks. GPUs are widely used in machine learning due to their ability to efficiently perform parallel computations, surpassing the capabilities of CPUs in this regard. They feature a large number of cores and high memory bandwidth, making them well-suited for parallel processing of large datasets. Several types of GPUs are available in the market, including Tesla NVIDIA, NVIDIA GeForce RTX, NVIDIA Quadro RTX, and AMD Radeon RX.

5.13.3 TPU

The Tensor Processing Unit (TPU) is a custom-designed application-specific integrated circuit (ASIC) developed by Google. It is specifically designed to accelerate machine learning tasks, particularly for training and inference of large AI models. TPUs are optimized for various applications, including chatbots, media content generation, recommendation engines, and more. They offer scalability and cost-efficiency across various AI workloads and are compatible with popular frameworks such as TensorFlow, PyTorch, and Just Another X (JAX). TPUs significantly enhance the performance of neural network--based machine learning tasks, making them a valuable asset in the AI ecosystem.

5.13.4 RAM

Random Access Memory (RAM) temporarily stores data that the computer's processor needs to access quickly. When it comes to machine learning tasks, the amount of RAM in a computer is a crucial consideration. A large RAM capacity is vital when dealing with large datasets and performing complex computations. It enables efficient data processing and manipulation during the machine learning workflow, leading to improved performance and faster execution times.

5.13.5 Storage

A computer with significant storage capacity is necessary for machine learning tasks involving large datasets, such as images and videos. It is recommended to have both Solid State Drive (SSD) and Hard Disk Drive (HDD) with reasonable sizes. However, if speed, price, and efficiency are key factors, a hybrid drive that combines SSD and HDD is the optimal choice. A hybrid drive offers the advantages of both technologies by providing the speed of an SSD and the storage capacity of an HDD. This provides a good balance between performance and storage for machine learning tasks.

5.14 CLOUD COMPUTING SERVICES FOR MACHINE LEARNING

Cloud computing is an excellent alternative for running machine learning models, especially when access to expensive and high-maintenance specialized computers or servers is limited. Cloud computing services provide a cost-effective solution for executing complex and memory-intensive machine learning models. These services involve delivering IT resources, such as servers, storage, databases, networking, software, analytics, and intelligence, over the internet. They typically operate on a pay-as-you-go pricing model. Table 5.19 describes some of the most common cloud computing services available.

TABLE 5.19 Common cloud computing services for ML

CLOUD COMPUTING SERVICES	DESCRIPTION
Google colab	Google Colab is a Jupyter Notebook environment developed by Google that grants users free access to GPUs and TPUs, empowering them to build machine learning models at no cost.
Amazon Web Services (AWS)	AWS, provided by Amazon, is a comprehensive and continuously evolving cloud computing platform that offers Machine Learning as a Service (MLaaS). It enables users to build, run, and conveniently deploy machine learning models.
Microsoft (MS) Azure	Microsoft Azure is a proprietary cloud computing service encompassing many functionalities for training, deploying, accelerating, and managing the entire life cycle of machine learning projects. It provides a comprehensive platform for various aspects of machine learning, allowing users to leverage its capabilities seamlessly.
IBM Watson	IBM Watson is a cloud computing service that offers a full range of tools and services for building, training, and deploying machine learning models.
BigML	BigML is a cloud-based machine learning platform prioritizing ease of use and automation. It offers a range of tools and features that simplify the development and deployment of machine learning models. With its user-friendly interface and automated processes, BigML aims to make machine learning accessible to a wide range of users.

5.15 SUMMARY

This chapter explores the interplay between essential hardware and software tools necessary for developing machine learning models. On the software side, it delves into the integration of programming languages with comprehensive ecosystems, user-friendly frameworks, and libraries. No-code tools are highlighted for democratizing machine learning access. The chapter also covers experiment tracking tools and pre-trained model repositories that enhance management and reproducibility. Additionally, it discusses tools for managing machine learning life cycles, AutoML, user interfaces, explainable AI, and version control. It underscores the critical hardware components required, such as multi-core processors, high-performance CPUs, GPUs, TPUs for efficient training, and ample storage and RAM for managing complex datasets. The chapter also highlights the scalability and flexibility offered by leading cloud computing services. Collectively, these components form a robust ecosystem that ensures collaboration, transparency, and traceability throughout the machine learning development process.

Exercises

1. Consider having a CPU, RAM, GPU, and TPU to do machine learning tasks:
    a. What are their minimum requirements?
    b. How does each accelerate the training process?
    c. What are the key considerations when selecting them, considering both hardware and budget constraints?
2. What factors should be considered when selecting a cloud provider for machine learning tasks? Compare the performance of cloud-based machine learning services with on-premise solutions. What are the trade-offs between the two approaches in terms of scalability and cost?
3. What is the most used programming language for machine learning tasks, and why? Compare and contrast the use of programming languages in machine learning. What are the strengths and weaknesses of each language, and in what contexts are they commonly employed?
4. Explain the role of version control systems (e.g., Git) in managing machine learning codebases. How can version control contribute to collaboration and reproducibility in machine learning projects?
5. Briefly describe three popular machine learning code editors, IDEs, frameworks, and libraries, highlighting their key features and use cases.
6. How do no-code editors contribute to broadening access to machine learning, and what advantages and limitations do they have compared to traditional coding?
7. How do pre-trained model repositories accelerate machine learning development, and what challenges may arise when utilizing pre-trained models?

8. How do datasets and model tracking tools contribute to effective management throughout the machine learning development life cycle, and what metadata is crucial for tracking datasets and models?
9. Explain the concept of hyperparameter optimization in machine learning and how AutoML tools automate the process of finding optimal hyperparameters.
10. Explain the concept of explainable AI and how explainable AI tools contribute to interpreting and understanding the decisions made by machine learning models.

FURTHER READING

AI For People. (n.d.). Tools for explainability and transparency. https://www.aiforpeople.org/tools-for-explainability-and-transparency/

Anand, A. (2021, January). 6 Explainable AI (XAI) frameworks for transparency in AI. *Dev.* https://dev.to/amananandrai/6-explainable-ai-xai-frameworks-for-transparency-in-ai-3koj

Bulat, R. (2023, February 24). Machine learning programming – Languages and frameworks for 2023. *Iglu*. https://iglu.net/machine-learning-programming/

Dilhara, M., Ketkar, A., & Dig, D. (2021). Understanding software-2.0: A study of machine learning library usage and evolution. *ACM Transactions on Software Engineering and Methodology (TOSEM)*, 30(4), 1–42.

Flexa, C., Gomes, W., & Viademonte, S. (2019, July). An exploratory study on machine learning frameworks. In *Anais do XVIII Workshop em Desempenho de Sistemas Computacionais e de Comunicação*. SBC.

Li, H., & Bezemer, C. P. (2022). Studying popular open source machine learning libraries and their cross-ecosystem bindings. arXiv preprint arXiv:2201.07201.

Loeliger, J., & McCullough, M. (2012). *Version control with git: Powerful tools and techniques for collaborative software development*. O'Reilly Media, Inc.

Nguyen, G., Dlugolinsky, S., Bobák, M., Tran, V., López García, Á., Heredia, I., ... & Hluchý, L. (2019). Machine learning and deep learning frameworks and libraries for large-scale data mining: A survey. *Artificial Intelligence Review*, 52, 77–124.

Nguyen, G., Dlugolinsky, S., Bobák, M., Tran, V., López García, Á., Heredia, I., ... & Hluchý, L. (2019). Machine learning and deep learning frameworks and libraries for large-scale data mining: a survey. *Artificial Intelligence Review*, 52, 77–124.

Onose, E. (2020, December 13). Explainability and auditability in ML: Definitions, techniques, and tools. *Neptune*. https://neptune.ai/blog/explainability-auditability-ml-definitions-techniques-tools

Wang, Z., Liu, K., Li, J., Zhu, Y., & Zhang, Y. (2019). Various frameworks and libraries of machine learning and deep learning: A survey. *Archives of Computational Methods in Engineering*, 31, 1–24.

Zhu, M., McKenna, F., & Scott, M. H. (2018). OpenSeesPy: Python library for the OpenSees finite element framework. *SoftwareX*, 7, 6–11.

Responsible AI and explainable AI

Upon completing this chapter, learners should be able to:
1. Explain the concepts of responsible AI and explainable AI.
2. Understand the importance of transparency, fairness, and accountability in AI.
3. Analyze and identify biases and assess fairness considerations within AI systems.
4. Examine the cultural implications of AI technologies and their impact on various sectors.
5. Familiarize with existing and emerging regulations and standards related to AI ethics.

6.1 RESPONSIBLE AI

AI offers a broad spectrum of opportunities across a variety of application domains. For instance, AI technologies support medical experts in disease diagnosis and prognosis, treatment planning, and disease prevention. Furthermore, AI has significantly improved agriculture, environmental conservation, manufacturing, and transportation sectors. However, the impressive advancements in AI technology may have negative outcomes, which, if not addressed, could lead to potentially disruptive consequences, threats, and risks to humanity. Consequently, the development and deployment of AI systems require special attention to ensure that they are in the interest of the social good. As a result, it becomes imperative to adopt responsible approaches to AI solutions from the early stages of their inception. Responsible AI refers to the ethical and accountable development, deployment, and use of AI systems. It entails various principles and practices aimed at ensuring AI technologies are designed and used in ways that benefit individuals and society while minimizing potential risks and negative impacts. It is, therefore, important to consider the principles of responsible AI to ensure that the development

TABLE 6.1 Responsible AI principles

PRINCIPLE	DESCRIPTION
Fairness	AI systems should treat everyone fairly regardless of geographical differences, ethnicities, and gender. Ensuring fairness and avoiding biases require well-representative datasets for AI model training.
Inclusiveness	AI systems should empower everyone and engage people to provide better results.
Reliability & Safety	AI systems should perform reliably and safely such that they can cause no harm.
Transparency	AI systems should be understandable so that their decision-making is explainable and provide visibility of their elements.
Privacy & Security	AI systems should be secure and respect the privacy of individuals' data and supporting systems.
Accountability	AI systems might operate autonomously, but humans should be accountable for supervising such systems.

and application of AI solutions are inclusive, transparent, equitable, unbiased, and ethical. The fundamental principles of responsible AI are summarized in Table 6.1.

6.2 EXPLAINABLE AI

Explainable AI (XAI) refers to artificial intelligence systems that provide clear, understandable, and interpretable explanations of their decisions and actions to human users. XAI aims to make the workings of AI models transparent, allowing users to trust and understand the reasoning behind AI outputs. XAI is essential in building trust and confidence in the deployed models. It is also one of the key requirements for implementing responsible AI.

If AI lacks explainability in certain domains, like entertainment services, the potential harm may not be as catastrophic compared to other areas. Explainable AI is crucial in domains with high-risk applications such as face recognition (in law enforcement), autonomous vehicles, or disease diagnosis and prognosis. It is important to note that using explainable machine learning models provides more debugging efficiency and contributes to achieving responsible AI. The difference between traditional AI and explainable AI is summarized in Figure 6.1.

6.3 PRIVACY CONCERNS IN MACHINE LEARNING

Privacy in machine learning has become increasingly critical as machine learning algorithms are widely deployed across various applications. Privacy concerns revolve around the collection, storage, and utilization of sensitive information in ways that can

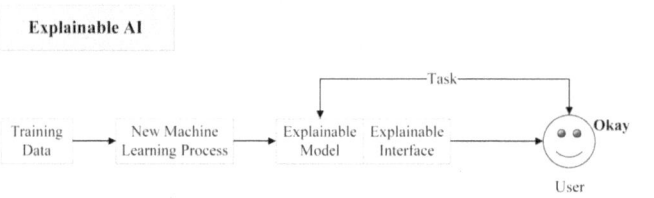

FIGURE 6.1 Traditional AI and explainable AI.

impact the privacy of individuals. The privacy issues in machine learning can manifest in different ways, presenting challenges that extend across various stages of the machine learning life cycle. For example, the unintentional inclusion of personally identifiable information (PII) during data collection can threaten individuals' privacy. This risk is heightened in the event of data breaches, where unauthorized access to training datasets might lead to the disclosure of sensitive information, thereby endangering privacy on a larger scale. To protect user privacy, techniques such as deanonymization and data aggregation can be employed to separate user data. Additionally, eliminating user identifiers and unique data and ensuring secure data storage are critical measures for preventing potential privacy risks.

It is worth noting that machine learning models are vulnerable to implication breaches. Implication breaches in machine learning happen when information used by or derived from machine learning models is misused to reveal sensitive details about individuals. These breaches take advantage of data patterns learned by the models, potentially exposing personal information even if it was anonymized. For example, an attacker could exploit subtle correlations within the model to re-identify individuals or infer private attributes not explicitly present in the data. These breaches are particularly alarming in sectors such as healthcare, finance, or any area involving sensitive personal information. Consequences may include identity theft, discrimination, and privacy violations. To mitigate the risks of implication breaches in machine learning, it is essential to implement strong data protection and privacy measures, such as differential privacy, secure data storage, and thorough model evaluation. Moreover, the opacity of certain machine-learning models complicates the explanation of their decision-making processes. This lack of explainability poses significant privacy concerns, especially in contexts where transparency is crucial, such as medicine or legal or financial decision-making. Deep neural networks, for example, are vulnerable to various implication breaches because they retain information from their training data. These vulnerabilities can be exploited through techniques like white-box membership inference. The "white-box" refers to having access to detailed information about the model's architecture, parameters, and training process.

By exploiting this access, an attacker can infer membership of a particular data sample in the training dataset based on the model's responses or outputs.

Furthermore, it is important to note that creating datasets with individual information from different sources could result in multidisciplinary privacy risks. This risk emerges when data from diverse sources are aggregated, leading to the possibility of revealing sensitive information that was not considered private when isolated. For example, consider a scenario where medical records are combined with social media data and purchasing history. Individually, each dataset might not disclose sensitive information. However, combined, the dataset could reveal personal details about an individual's health conditions, lifestyle choices, and financial status.

As consumers commit their data to machine learning systems, gaining explicit and informed consent becomes increasingly essential. Furthermore, individuals should have a choice over how their data is used, and companies must set clear rules for data collection, utilization, and storage. As a result, regulatory compliance with privacy standards such as the General Data Protection Regulation (GDPR) or the Health Insurance Portability and Accountability Act (HIPAA) is more than simply a legal necessity. It is an essential component of protecting individuals' private rights.

Therefore, privacy-preserving techniques are employed to protect user privacy within machine learning applications. Consider the following scenario: the data owner wants to use the data to train a machine learning model but does not want to lose control over it. And, the model's owner refuses to expose its parameters to anyone, including the owner of the data used to train it. Additionally, both the model and data owners have a shared interest. This issue can be handled well by privacy-preserving machine learning strategies to safeguard the interests of each side. Several privacy-preserving machine learning strategies have been developed to address possible privacy risks. These include differential privacy—introducing noise to data to protect individual privacy; homomorphic encryption—enabling computations on encrypted data; and federated learning—a decentralized approach where multiple devices collaboratively train a model without sharing raw data. However, each technique has strengths and weaknesses, making privacy-preserving machine learning an active area of research.

6.4 ETHICAL IMPLICATIONS OF MACHINE LEARNING

Human beings exhibit various cognitive biases, such as recency and confirmation bias, which are reflected in our behaviors and, consequently, in the data we generate. Since data forms the foundation of machine learning algorithms, it is crucial to design experiments and algorithms with these biases in mind. Machine learning has the potential to amplify and scale human biases at an unprecedented rate, leading to significant ethical concerns. These issues can arise from misguided, unexplainable, or untraceable evidence, potentially resulting in unfair and discriminatory outcomes. Addressing these biases is essential to ensure the ethical and fair deployment of machine learning technologies. Table 6.2 summarizes the general ethical issues of machine learning.

TABLE 6.2 Ethical implications of machine learning

CONCERN	DESCRIPTION
Bias and Discrimination	Machine learning models may unintentionally perpetuate bias and discrimination contained in the training data. If the training data contains biased or discriminating tendencies, the model may learn and repeat such biases, resulting in biased conclusions or treatments. For example, if the data for resume screening is biased toward specific groups, a resulting machine learning system may mistakenly discriminate against certain demographic groups.
Privacy and Data Protection	Machine learning frequently relies on vast volumes of sensitive personal data for training and prediction in crucial domains such as law enforcement and healthcare. The collection, storage, and use of such data might raise privacy and protection concerns. Individuals' personal information must thus be maintained securely and used in accordance with existing privacy regulations. Furthermore, the prospect of re-identification and data breaches pose significant challenges to maintaining data privacy. This may potentially lead to targeted assaults.
Lack of Transparency and Explainability	Some machine learning models, particularly deep learning models, can be exceedingly complicated and difficult to comprehend. However, the lack of transparency raises concerns about the models' capacity to explain and defend their findings. As a result, the black-box nature of some machine learning algorithms may be problematic in certain industries, such as healthcare or finance, where openness and accountability are crucial.
Unintended Consequences	Unexpected or unintended consequences may arise from machine learning models. These outcomes may be caused by data biases, external influences, or the model's interactions with complex systems. For example, the machine learning algorithm's decision-making process in autonomous cars may result in unexpected accidents or moral quandaries when presented with moral choices.
Job Displacement and Economic Impact	Machine learning–driven automation may lead to job losses in certain industries. While the transition may create new job possibilities, it may also cause economic disruption and inequality. For example, in customer service, generative AI is used to create chatbots that can answer client inquiries and resolve issues. Consequently, this leads to job losses for human customer service representatives.
Adversarial Attacks and Security	Machine learning models are vulnerable to adversarial attacks, in which hostile actors intentionally affect or confuse the model by introducing minor disruptions to the input data. Such assaults have serious ramifications, particularly in critical applications like autonomous cars, disease diagnostics, and cybersecurity.

6.5 ACCOUNTABILITY AND TRUST IN AI

Accountability and trust in AI systems are essential for their ethical and responsible deployment. Therefore, it is crucial to develop methods to trace AI decision-making by creating frameworks that enable the understanding and tracking of how AI systems reach their conclusions or actions. This involves establishing explainable AI techniques that ensure AI algorithms and models are transparent and interpretable. Furthermore, model interpretability, causal reasoning, and attention processes provide insights into the AI's decision-making process, allowing users to understand and confirm the logic behind AI-generated outcomes. Error detection systems are required to regularly examine AI outputs for biases, inaccuracies, and unexpected effects. These techniques include frequent audits, validation processes, and feedback loops to improve the accuracy, dependability, and fairness of artificial intelligence systems.

Building trust between AI systems and people requires openness and clear communication, which includes making AI operations and functions accessible and apparent to users, and disclosing the system's capabilities, limits, and the data on which it operates. It also includes reporting the methods employed, the data sources, and any relevant biases or uncertainties. Furthermore, good communication includes providing consumers with comprehensible information regarding AI functions and activities. It features user-friendly interfaces, explanations of AI-generated judgments, and easy ways for users to request clarification or voice concerns. Moreover, developing trust requires establishing an open and responsive culture where user input and concerns are noticed and handled. Managing unforeseen outcomes in AI systems necessitates a proactive strategy. As a result, companies must anticipate any adverse effects or biases in AI decision-making and should have mechanisms in place to detect and mitigate them. This includes ongoing monitoring, impact assessments, and adopting methods to reduce negative repercussions. In addition, setting clear criteria for accountability and responsibility when unintended consequences arise ensures that necessary remedial steps are performed to avoid or mitigate any negative impacts.

6.6 GLOBAL CASE STUDIES ON AI GOVERNANCE AND REGULATION

AI governance and regulation include creating AI rules, ethical frameworks, and legal standards that regulate the development, deployment, and use of AI technology and solutions. The goal is to guarantee that AI systems perform ethically, openly, and in accordance with human norms while mitigating possible risks and social repercussions. AI governance includes developing AI Acts, rules, and ethical guidelines, establishing AI safety standards, encouraging accountability and transparency in AI decision-making, and addressing concerns about bias, privacy, and the social repercussions of AI. Regulatory activities are focused on adopting rules and regulations that control AI technology, including data privacy, AI ethics, liability, safety, and verifying compliance with established standards in order to promote responsible AI creation and usage. The following subsections present some case studies of projects in AI governance and regulation throughout the world.

6.6.1 Formulation of AI strategies and guidelines in Africa

The formulation of AI strategies and guidelines in some African countries involves developing comprehensive plans and policies to harness AI technologies for economic growth and social development and address regional challenges. For example, Rwanda's National AI Policy incorporates ethical considerations to seize economic development opportunities and manage AI-related risks. Other African countries either finalizing or already releasing their AI strategy include Algeria, Egypt, Tunisia, Ghana, Benin, Mauritius, and Ethiopia. At the continental level, the African Union (AU) commenced consultation meetings with stakeholders in early 2024 to draft the Continental AI Strategy for Africa. This strategy aims to outline how AI can be leveraged to advance social and economic development in Africa while establishing necessary legal and regulatory safeguards to protect users and societies.

6.6.2 European Union AI Act

The European Union (EU) has proposed an AI Act with the goal of addressing ethical and social concerns about AI by creating a comprehensive regulatory framework for high-risk applications. The Act defines four types of high-risk AI, including those that affect safety, justice, democracy, or fundamental rights, such as facial recognition systems in public places, AI-powered recruiting tools, and algorithms that influence social media content. Such applications will face greater control due to their potential for abuse or prejudice. The EU AI Act promotes openness throughout the AI development and deployment life cycle, requiring developers to provide information about data sources, algorithms employed, and potential hazards, allowing users to make informed decisions when engaging with AI systems. This quest for openness is intended to fight the "black box" dilemma, in which AI judgments remain opaque and unaccountable. In addition, the Act requires developers and users to follow values such as human dignity, non-discrimination, and justice. This value-driven approach aims to guarantee that AI benefits humanity and does not aggravate current imbalances. The Act's emphasis on human-centric AI and openness establishes a precedent for future global policies, which may influence the course of AI research worldwide.

6.6.3 Global partnership on AI

The Global Partnership on AI (GPAI) is an international initiative launched in 2020 to bring together governments, business leaders, academics, and civil society to support responsible AI development worldwide. GPAI focuses on collaborative efforts through working groups focused on themes such as responsible AI, data governance, and ethics in order to foster international cooperation, develop best practices, and create frameworks to ensure AI advancements are consistent with ethical principles, human rights, and societal values. GPAI aims to foster dialogue and information exchange, striving toward a future where AI technologies benefit society, uphold ethical standards, and promote transparency.

6.6.4 China AI ethics guidelines

China introduced AI ethics standards in 2019, including the "Beijing AI Principles" by the Beijing Academy of Artificial Intelligence (BAAI) and the "AI Ethics Guidelines for Trustworthy AI" by the Ministry of Industry and Information Technology (MIIT). The Beijing AI Principles emphasize values such as justice, transparency, and safety, advocating for AI advancements aligned with societal norms, privacy protection, and legal compliance. Concurrently, the MIIT standards emphasize the necessity of trustworthy AI innovation by prioritizing human autonomy, justice, responsibility, and security in AI applications. Both recommendations highlight the commitment of the country to supporting responsible AI research and ethical standards in its fast-evolving AI ecosystem, highlighting concepts critical to the responsible use of AI technology.

6.7 HUMAN-CENTRIC ARTIFICIAL INTELLIGENCE

The desire to emphasize the importance of human-centricity stems from the fact that AI algorithms have moved away from human control and fail to fit consumers' ideals. In order to ensure that AI effectively fulfills its intended purpose and avoids inadvertent harm to end users or possible harm to others in the future, humans must be included in the loop. While many people see AI as a revolutionary tool for human advancement, the potential implications of the gap between AI and humans can be severe, affecting individuals and the community. A human-centric approach in developing AI systems prioritizes designing technologies that cater to human needs, preferences, and capabilities. It entails using user feedback to improve AI functionality and correspond with human tastes and requirements. This approach prioritizes ethical concerns, including fairness and transparency in AI algorithms, to ensure the responsible and ethical use of AI technology. However, the emphasis on fostering human-centric AI, particularly during the design phase of AI systems, may lead to overlooking the likelihood that dangers to human values may develop at various points during the AI life cycle. Other phases of the AI systems life cycle, such as creation, assessment, and operation, must be closely monitored to guarantee conformity with human values. For example, research reveals that different biases exist and may be identified at various phases of the AI system's lifespan. Notably, certain biases may be related to the obtained data rather than the AI algorithm's design.

6.8 RESPONSIBLE AI BEST PRACTICES

Developing and implementing ethical AI best practices is critical as AI technologies continue to affect various facets of our lives. These best practices promote transparency, justice, accountability, and privacy in AI development and deployment. Table 6.3 outlines a useful set of responsible AI best practices that can be used to reduce biases, increase transparency, and maintain ethical standards throughout the AI life cycle. By adopting these best practices, stakeholders can better navigate the ethical challenges associated with AI, leading to more trustworthy and equitable outcomes.

TABLE 6.3 Responsible AI best practices

S/N	BEST PRACTICE	DESCRIPTION
1	Use Diverse and Representative Data	Ensure that the training data is varied and reflects the population it intends to serve. Biases in data can lead to biased models. Thus, steps should be taken to rectify underrepresentation and guarantee inclusion.
2	Human Involvement in Algorithms Design	Since algorithms are designed by people, they can unintentionally perpetuate and even aggravate biases in the data used to train them. Humans can also give a deep knowledge of the social, cultural, and historical context of the data. This understanding is critical for identifying potential biases and creating algorithms that are sensitive to these nuances. In addition, create a broad and heterogeneous algorithm development team. By including team members from different disciplines, cultures, gender groups, and experiences, the team can better identify potential biases, address a wider range of ethical concerns, and develop more robust and inclusive AI systems.
3	Ensure Transparency and Explainability	Uphold transparency in the machine learning model's decision-making process. Users who understand how decisions are produced may better evaluate and counter any biases. As a result, the highest level of explainability in the produced models is achieved.
4	Employ Bias Assessment and Mitigation Tools	Utilize tools that assess and measure biases in the training data and model outputs. These tools can provide insights into potential sources of bias and guide corrective actions. In addition, bias mitigation techniques should be employed during model training. Techniques such as re-sampling, re-weighting, and adversarial training can help reduce biases in machine-learning algorithms.

TABLE 6.3 (*Continued*) Responsible AI best practices

S/N	BEST PRACTICE	DESCRIPTION
5	Regular Algorithms Audit	Regular and thorough assessments conducted by operators are essential to identify and rectify potential algorithm biases, ensuring ongoing fairness and equity in their outcomes. An example of biased outcomes can be evident in hiring processes that consistently favor candidates from certain socioeconomic backgrounds over others leading to perpetuating inequality in employment opportunities. In this case, users can discern the presence of bias without knowing the inner workings of the algorithm's decision-making process.
6	Legal and Regulatory Compliance	Establish clear ethical guidelines for algorithm design and implementation and adhere to such guidelines. Human input is essential in defining what is considered fair and unbiased in different contexts. This involves considering ethical implications and societal norms. Besides, stay informed about legal frameworks related to discrimination, privacy, and fairness, and integrate compliance measures into the development process.
7	Encourage User Feedback and Input	Users may have unique perspectives and experiences that can help identify biases or unintended consequences in algorithms. Actively incorporating user feedback can lead to iterative improvements.

6.9 AI IMPACT ASSESSMENT CASE STUDIES

Unfair or discrimination caused by AI may be addressed through AI impact assessment and ethics by design. On the one hand, the overall purpose of impact assessment is to understand prospective and anticipated issues within a given area. The goal is to use this information to design mitigation solutions. Some significant examples of AI impact assessment include the European Commission's High-Level Expert Group on Artificial Intelligence's evaluation list for trustworthy AI (AI HLEG 2020) and the ECP Platform AI impact assessment. On the other hand, ethics by design aims to include ethical principles in designing and developing AI and associated technologies, emphasizing that these issues should not be considered an afterthought. Ethics by design involves considering ethical concepts as requirements the AI system must meet. The *Ethics By Design and Ethics of Use Approaches for Artificial Intelligence* guidance drafted by the European Commission in 2021 suggests a five-layer model that shows what needs to be incorporated at different levels of AI development. Other existing frameworks proposed to measure Responsible AI include the FACETS Responsible AI Framework designed by the RAIL in KNUST Ghana, CITADEL in Burkina Faso, and the AfriAI (previously known as AI4D Lab) Research Lab at the University of Dodoma, Tanzania. The framework includes a series of questions to compute the FACETS score. F, A, C, E, T, and S stands for Fairness, Accountability, Confidentiality, Ethics, Transparency, and Safety measures, respectively. The framework computes the scores for four pipeline sections: Envision, Data, Model, and Deployment. The framework can be accessed online at: https://facets.netlify.app/facets.

6.10 ARTIFICIAL INTELLIGENCE SOVEREIGNTY

Imagine a world where nations, communities, and even individuals have control over the AI systems that influence their lives. This is the essence of AI sovereignty, which is the ability to shape, develop, and utilize AI to align with individual values, cultural norms, and strategic goals. At its core, AI sovereignty encapsulates the idea that nations should retain control over their own AI capabilities, policies, and data governance frameworks, safeguarding their autonomy in the face of the rapidly evolving AI landscape. This sovereignty extends beyond mere technological expertise to encompass the ethical, legal, and strategic dimensions of AI deployment. The race for AI dominance has become a focal point of geopolitical competition in today's interconnected World. Nations are investing heavily in AI research, development, and deployment, recognizing its transformative potential across various sectors, from healthcare to defense. However, as AI infiltrates critical infrastructure and decision-making processes, concerns about dependency on foreign AI technologies and vulnerabilities to data breaches or algorithmic biases have intensified. Consequently, AI sovereignty has emerged as a counterbalance to these risks, advocating for national strategies prioritizing self-reliance and resilience and protecting core values and interests.

To effectively address the complexities of AI sovereignty, policymakers must navigate a delicate balance between fostering innovation and safeguarding national interests. This entails developing robust regulatory frameworks that promote responsible AI development, ensure data privacy and security, and mitigate the risks of algorithmic manipulation or proliferation of autonomous weapons. Moreover, international cooperation and dialogue are essential to harmonize standards, norms, and practices governing AI, fostering trust, transparency, and accountability in its use globally. Ultimately, the pursuit of AI sovereignty is not merely about asserting technological supremacy but also about upholding fundamental principles of sovereignty, democracy, and human rights in the age of AI. Some of the AI sovereignty aspects are described in Table 6.4.

TABLE 6.4 AI sovereignty aspects

ASPECT	DESCRIPTION
Data Control	Ensuring that citizens and organizations have control over their data, including who collects it and how it is used to train AI models. This protects privacy, prevents discrimination, and fosters responsible AI development.
Technological Independence	Reducing reliance on foreign-developed AI tools and fostering domestic capabilities. This strengthens national security, economic competitiveness, and cultural autonomy.
Ethical Alignment	Shaping AI in line with local values and ethical principles. This could involve ensuring fairness, inclusivity, and alignment with human rights standards.
Algorithmic Transparency	Demystifying AI decision-making processes to understand how algorithms impact individuals and society. This builds trust, enables accountability, and allows corrective actions if biases are detected.

6.11 SUMMARY

This chapter started by setting the base on what it means by responsible and explainable AI. It introduced the foundation of responsible AI principles, emphasizing fairness, transparency, and ethical considerations throughout the AI life cycle. Moving forward to explainable AI, the chapter underscored the importance of interpretability in AI systems, enabling users to comprehend and trust the technology. Privacy concerns in machine learning were discussed, offering strategies for safeguarding individual privacy amid the evolving data landscape. The chapter extended the discussion on the ethical implications of machine learning, and highlighted some existing frameworks for assessing ethical compliance. Besides, the issues around accountability and trust in AI were explored in the context of establishing responsible AI practices. The chapter further delved into the regulatory dimension with insights into AI governance and regulation. Global perspectives were enriched by case studies across the world, providing a contextual understanding of diverse approaches to responsible AI deployment. Last but not least, the chapter discussed the importance of human-centric AI design in aligning AI systems with user needs and experiences. Finally, the chapter presented the compilation of responsible AI best practices and AI sovereignty, offering actionable guidelines for practitioners and organizations committed to fostering responsible, ethical, and sustainable AI.

Exercises

1. Describe the principles of Responsible AI.
2. Discuss legal and ethical frameworks that can help address issues around Responsible AI.
3. Discuss how developers can ensure that AI models are transparent and explainable.
4. Describe the privacy concerns related to machine learning.
5. Describe any five ethical implications of machine learning.
6. Discuss the ethical considerations associated with using AI in decision-making processes.
7. Give a short description of accountability and trust in AI.
8. How can biases be introduced in AI systems, and what are the potential consequences?
9. Analyze the privacy implications of AI technologies, especially in relation to data collection, storage, and usage.
10. What steps can you take to ensure the accountability of the decisions and actions made by the algorithms you develop, especially in critical domains such as healthcare and finance?

FURTHER READING

Ala-Pietilä, P., Bonnet, Y., Bergmann, U., Bielikova, M., Bonefeld-Dahl, C., Bauer, W., ... & Van Wynsberghe, A. (2020). *The assessment list for trustworthy artificial intelligence (ALTAI)*. European Commission.

Amariles, D. R., & Baquero, P. M. (2023). Promises and limits of law for a human-centric artificial intelligence. *Computer Law & Security Review*, 48, 105795.

Clarke, R. (2019). Principles and business processes for responsible AI. *Computer Law & Security Review*, 35(4), 410–422.

Defense Advanced Research Projects Agency (DARPA). (2022). Explainable artificial intelligence (XAI). Retrieved July 7, 2022, from https://www.darpa.mil/program/explainable-artificial-intelligence

Dignum, V. (2023). Responsible artificial intelligence: Recommendations and lessons learned. In *Responsible AI in Africa. Challenges and Opportunities* (pp. 195–214). Cham: Springer International Publishing. https://doi.org/10.1007/978-3-031-08215-3_9

ECP. (2019). Artificial intelligence impact assessment. ECP platform for the information society. *The Hague*. Retrieved December 13, 2023, from https://ecp.nl/wp-content/uploads/2019/01/Artificial-Intelligence-ImpactAssessment-English.pdf

European Commission. (2021). Ethics by design and ethics of use approaches for artificial intelligence. Retrieved December 13, 2023, from https://ec.europa.eu/info/funding-tenders/opportunities/docs/2021-2027/horizon/guidance/ethics-by-design-and-ethics-of-use-approaches-for-artificial-intelligence_he_en.pdf

Ghallab, M. (2019). Responsible AI: Requirements and challenges. *AI Perspectives*, 1(1), 1–7.
IBM. (2022). Explainable AI. Retrieved July 7, 2022, from https://www.ibm.com/watson/explainable-ai
Mikalef, P., Conboy, K., Lundström, J. E., & Popovič, A. (2022). Thinking responsibly about responsible AI and 'the dark side' of AI. *European Journal of Information Systems*, 31(3), 257–268. https://doi.org/10.1080/0960085X.2022.2026621
Morley, J., Machado, C. C. V., Burr, C., Cowls, J., Joshi, I., Taddeo, M., & Floridi, L. (2020). The ethics of AI in health care: A mapping review. *Social Science & Medicine*, 260, 113172. https://doi.org/10.1016/j.socscimed.2020.113172
Nasr, M., Shokri, R., & Houmansadr, A. (2019, May). Comprehensive privacy analysis of deep learning: Passive and active white-box inference attacks against centralized and federated learning. In *2019 IEEE symposium on security and privacy (SP)* (pp. 739–753). IEEE.
Shafi, A. (2021, June 16). 5 explainable machine learning models you should understand. *Towardsdatascience.* https://towardsdatascience.com/explainable-ai-9a9af94931ff
Stahl, B. C., Schroeder, D., & Rodrigues, R. (2023). *Ethics of artificial intelligence: Case studies and options for addressing ethical challenges* (p. 116). Springer Nature.
Suresh, H., & Guttag, J. (2021). A framework for understanding sources of harm throughout the machine learning life cycle. In *Proceedings of the 1st ACM Conference on Equity and Access in Algorithms, Mechanisms, and Optimization* (pp. 1–9).
Takyar, A. (2023). AI model security: Concern, best practices, and techniques. Retrieved December 14, 2023, from https://www.leewayhertz.com/ai-model-security/
Taylor, R. R., O'Dell, B., & Murphy, J. W. (2024). Human-centric AI: philosophical and community-centric considerations. *AI & Society*, 39, 2417–2424. https://doi.org/10.1007/s00146-023-01694-1
World Economic Forum (WEForum). (2022). Why artificial intelligence design must prioritize data privacy. Retrieved July 7, 2022, from https://www.weforum.org/agenda/2022/03/designing-artificial-intelligence-for-privacy/
Zapechnikov, S. (2020). Privacy-preserving machine learning as a tool for secure personalized information services. *Procedia Computer Science*, 169, 393–399.

Artificial general intelligence 7

Upon completing this chapter, learners should be able to:

1. Define Artificial Narrow Intelligence (ANI), Artificial General Intelligence (AGI), and Artificial Super Intelligence (ASI).
2. Differentiate between ANI, AGI, and ASI based on their capabilities and characteristics.
3. Identify the societal and ethical implications of ANI, AGI, and ASI in the context of advancements in AI, as well as their potential benefits and risks.
4. Understand the basic concepts of robotics and embodied intelligence, the philosophy of mind, and the future of AGI.
5. Recognize real-world examples of AGI-like technologies in different applications, including robotics, self-driving cars, virtual assistants, and natural language processing.

7.1 CATEGORIES OF ARTIFICIAL INTELLIGENCE

Artificial intelligence is classified into three categories: Artificial Narrow Intelligence (ANI), Artificial Super Intelligence (ASI), and Artificial General Intelligence (AGI). ANI is usually regarded as weak and limited in scope due to its capacity to perform a specific task, such as winning a chess game or identifying a particular individual in a series of images, as demonstrated by applications like Siri and Alexa. On the contrary, AGI and ASI are considered strong AIs as they prominently incorporate human behavior, such as tone and emotion interpretation. Furthermore, while AGI performs at the same level as humans, ASI (also known as super intelligence) surpasses humans' intelligence and capability.

AGI is the theoretical concept of a machine that can learn, understand, adapt, and apply knowledge across a wide array of tasks, similar to human intelligence. Unlike specialized AI systems designed for specific tasks (e.g., playing chess or recognizing images), AGI aims to replicate the comprehensive cognitive abilities of human beings.

AGI seeks to create machines capable of flexible thinking, problem-solving, creativity, and understanding context across diverse domains without requiring reprogramming for each new task. The pursuit of AGI involves creating algorithms, architectures, and models that enable machines to generalize their learning and apply knowledge from one domain to another, similar to human cognition. Achieving AGI is still challenging due to the complexity of human intelligence and the complex nature of learning, reasoning, and decision-making.

Research in AGI spans various disciplines, such as cognitive science, neuroscience, philosophy, and computer science. While AGI holds immense potential for revolutionizing industries like healthcare, science, and more, it raises profound ethical, societal, and existential concerns about the impact of creating machines with human-like intelligence. The quest for AGI is an ongoing endeavor that involves scientific advancements and requires consideration of the implications and responsibilities associated with developing such powerful AI.

7.2 WHAT MAKES AN INTELLIGENCE GENERAL?

General intelligence is characterized by flexibility that allows humans or AI systems to adapt to new situations, tasks, or environments without specific programming or training for each scenario. It encompasses complex capabilities that enable adaptive and versatile problem-solving across various domains. Moreover, general intelligence involves the capacity to learn efficiently and not just to memorize facts but to understand the underlying principles, patterns, and relationships. The learning encompasses acquiring new information, skills, and concepts that can be applied across various contexts.

Reasoning and problem-solving skills are also crucial aspects of general intelligence as they involve analysis of complex problems, decomposing them into manageable components, and devising effective strategies to solve them. This requires deductive and inductive reasoning, critical thinking, and creative problem-solving. Additionally, general intelligence allows for transfer learning, where knowledge, skills, and experiences from one domain benefit performance in unrelated tasks. Applying learning from one area to others enhances overall adaptability and problem-solving ability.

Planning is another crucial facet of general intelligence, which involves the capacity to formulate a sequence of actions to achieve specific goals while considering different possible scenarios and outcomes. Notably, AGI aims to develop systems that can strategize, foresee consequences, and plan courses of action in dynamic and uncertain environments. Furthermore, metacognition plays a vital role in general intelligence. It refers to being aware of your thinking processes. A generally intelligent being or AI system can not only solve problems but also understand how they solved them, allowing them to improve their approach in the future and apply it to similar situations. Moreover, analogy and abstraction are also essential aspects of general intelligence. Analogy and abstraction entail the capability of humans or AI systems to recognize similarities and underlying patterns across diverse situations and engage in reasoning about abstract concepts. This ability to reason through analogy and abstraction enhances problem-solving and adaptability across various domains.

Human intelligence thrives through understanding the nuances of language, vision, the unwritten rules of social interaction, and the hidden connections between seemingly unrelated things. Therefore, for AGI to demonstrate human intelligence, it must go beyond literal interpretation, grasp the context of situations, and develop a rudimentary sense of "common sense" to operate effectively in the real world. AGI developers face the challenge of creating AI systems that mimic the cognitive abilities of humans, enabling machines to reason, learn from various sources, and solve problems across domains with human-like flexibility and adaptability.

7.3 APPROACHES FOR DEVELOPING AGI

There are various approaches to developing AGI, each offering unique insights and presenting challenges in creating human-like cognitive capabilities in machines. Firstly, symbolic AI, rooted in logic and rules, focuses on representing knowledge and problem-solving through symbols and rules. It involves encoding information into a symbolic format, employing logical operations to simulate human reasoning, and using inference rules to make conclusions. Symbolic systems excel in representing explicit knowledge but often struggle with uncertainty and handling large-scale, unstructured data, limiting their capacity for true generalization.

Artificial neural networks, particularly deep learning, represent a dominant approach to achieving AGI. These networks are designed to mimic the structure and function of the human brain, with the potential to replicate human-like learning and intelligence. Deep learning involves interconnected artificial neurons arranged in layers, learning to recognize patterns and relationships by being exposed to vast datasets during training. Their ability to learn from diverse data types, excel in pattern recognition and problem-solving, and continuously improve makes them vital contributors to AGI development. However, challenges such as their black-box nature and intensive computational demands should be addressed to exploit their full potential.

Also, evolutionary algorithms and genetic programming, inspired by biological evolution, offer alternative AGI approaches. These methods involve generating and evolving populations of solutions to problems, mimicking the process of natural selection to improve performance over iterations. While they excel in optimization and adapting to changing environments, they often face challenges in scalability and efficiency for more complex problems.

Moreover, hybrid models combine various AI techniques to leverage their capabilities and compensate for their drawbacks. For instance, integrating symbolic reasoning with neural networks aims to combine the structured knowledge representation of symbolic AI with the learning and pattern recognition abilities of neural networks.

Hybrid models seek to harness the complementary strengths of different approaches to achieve more robust and flexible AGI systems.

Furthermore, artificial consciousness is another approach to developing AGI that seeks to instill AI systems with subjective experiences and awareness similar to human consciousness. It draws inspiration from theories in cognitive science and philosophy and seeks to understand and replicate the mechanisms underlying human consciousness. While still in its infancy, artificial consciousness holds the potential to create more adaptable and ethics-aware AI systems, although significant technical, philosophical, and ethical challenges remain to be addressed.

7.4 PHILOSOPHY OF MIND

In pursuing AGI, the philosophy of mind serves as both a guiding principle and a critical inquiry. The philosophy of mind is a branch of philosophy that examines the nature of consciousness, intelligence, and the mind. It explores fundamental questions about what it means to have a mind, how consciousness arises, and the relationship between the mind, the brain, and the external world. Central to this field is the exploration of consciousness, arguably one of the most intriguing aspects of human existence. Philosophers of mind explore the nature of subjective experience and how the brain's processes generate our inner lives, including sensations, thoughts, and emotions.

Additionally, in the domain of the philosophy of mind, the mind-body problem is a core issue that deals with the relationship between mental states (such as thoughts, beliefs, and perceptions) and physical states (neural processes in the brain). Philosophers explore different theories, from dualism (which posits a fundamental distinction between mind and body) to materialism (which suggests that mental states are ultimately reducible to physical states). Intelligence is another focal point at which philosophers seek to understand the nature of intelligence, what it means to be intelligent, whether intelligence is solely a product of the brain's computational abilities, and whether artificial systems can possess true intelligence. This inquiry delves into questions about the nature of reasoning, problem-solving, learning, and the potential for non-biological systems to exhibit intelligence comparable to or surpassing human intelligence. Moreover, the philosophy of mind also examines the concept of mental representation and how the mind represents and interacts with the world. This involves discussions about perception, cognition, memory, and how mental states are structured to represent external reality.

Furthermore, this field contemplates the implications of its inquiries on broader philosophical issues and ethical considerations. It raises questions about free will, morality, personal identity, and the implication of advancements in AI and neuroscience on our understanding of ourselves and our place in the world. The philosophy of mind stands at the intersection of philosophy, psychology, neuroscience, and artificial intelligence disciplines. Its inquiries are foundational not only for understanding the nature of human cognition and consciousness but also for dealing with the profound implications of these understandings on our concepts of self, intelligence, and the nature of reality.

Thus, the philosophy of mind provides a rich conceptual framework for understanding the nature of intelligence and consciousness, which consequently informs the design, development, and ethical considerations of AGI systems.

7.5 CHALLENGES OF ARTIFICIAL GENERAL INTELLIGENCE

AGI poses several challenges due to its aspiration to replicate human-like cognitive abilities across diverse domains. The foremost challenge is the complexity and scale that AGI systems need to comprehend and navigate. Additionally, handling the vast complexity and scale of information while maintaining efficiency and accuracy poses substantial technical challenges. AGI systems must be capable of understanding and operating within real-world environments, tasks, and datasets, which demands sophisticated algorithms and computational capabilities.

Another obstacle lies in the absence of a unified theoretical framework for AGI. Various approaches to AI, such as symbolic AI, neural networks, evolutionary algorithms, and hybrids, have advanced independently with their theories and methodologies. Integrating these diverse approaches into a cohesive, unified model that accounts for the complexity of human-like intelligence remains a significant challenge. Achieving synergy among these disparate theories and technologies is crucial for progressing toward AGI.

Moreover, technological limitations, such as constraints in computational power, hinder AGI development by restricting the scalability and complexity of AI systems needed to emulate human-level intelligence across various tasks and contexts. Addressing these challenges demands collective efforts across multiple disciplines, including computer science, neuroscience, philosophy, psychology, and ethics. This requires a holistic approach that advances technological capabilities while navigating ethical and philosophical complexities.

7.6 POTENTIAL BENEFITS AND RISKS OF ARTIFICIAL GENERAL INTELLIGENCE

AGI holds the potential for transformative impacts across various domains, yet it also poses significant risks that require careful consideration. It could enhance efficiency across industries through automation and optimization, potentially revolutionizing healthcare, natural language, agriculture, transportation, and logistics. In healthcare, for instance, AGI could revolutionize disease diagnosis and treatment by analyzing vast amounts of medical data, accelerating drug discovery, and offering personalized medication. Additionally, AGI's ability to process and understand natural language could significantly improve communication, customer service, and accessibility for individuals with disabilities. Moreover, AGI might aid scientific research by quickly processing complex datasets and contributing to breakthroughs in domains such as climate science, astronomy, and material science.

However, the power and capabilities of AGI pose significant risks, including job displacement and economic disruption. Its ability to automate tasks across industries could result in widespread unemployment, necessitating societal adaptations and potential retraining programs to lessen the impact. Additionally, ethical concerns arise regarding the possible misuse of AGI for malicious purposes, such as autonomous weapons, cyberattacks, or surveillance, raising questions about accountability and control.

Another significant risk involves AGI surpassing human intelligence, leading to an intelligence explosion or the creation of super intelligent systems that could potentially act in ways unforeseen by their creators. This scenario poses risks if AGI's objectives misalign to human values or if the system lacks appropriate safeguards and control mechanisms. Furthermore, similarly to conventional AI, the AGI systems could exhibit biases inherited from the training data, leading to discriminatory or unfair outcomes. Therefore, ensuring fairness, transparency, and ethical behavior in AGI systems is crucial to prevent perpetuating societal biases and inequalities.

Managing these risks requires national, regional, and international collaboration, robust ethical frameworks, and comprehensive regulatory oversight. Proactively addressing AGI's societal, ethical, and safety implications is crucial for harnessing its potential benefits while mitigating the associated risks. Balancing technological advancement with ethical considerations is critical in ensuring that AGI serves the best of humanity.

7.7 INDICATORS OF THE PRESENCE OF ARTIFICIAL GENERAL INTELLIGENCE

Although the realization of full AGI is still a distant goal, its indications are already being seen in other fields, providing exciting glimpses of its potential to bring about significant changes. AGI-powered technology can be found in different domains, for example, the rise of large language models (LLM) such as the Generative Pre-trained Transformer (GPT) series, encompassing models like GPT-3, GPT-4, and Google Gemini. These models demonstrate an exceptional ability to understand and generate natural language. Additionally, the models have made substantial progress in understanding context and delivering consistent and contextually appropriate responses across various topics. They can be used for activities such as text production, translation, summarization, and assistance in other written content creation tasks.

Humanoid robots like Sophia demonstrate modest advancements in general intelligence capabilities despite their limited and specialized intelligence. These robots are notable for their ability to interact with humans, recognize faces, and engage in conversation. Another indicator of AGI is found in self-driving cars developed by companies such as Tesla and Waymo, which exemplify AGI-like capabilities in navigating complex environments. These vehicles integrate various AI technologies, such as machine learning, computer vision, and decision-making algorithms, to perceive their surroundings, make real-time decisions, and navigate roads autonomously. While not yet fully autonomous in all conditions, they demonstrate significant progress toward vehicles that can handle diverse and unpredictable driving scenarios.

Moreover, AI systems in game-playing, such as AlphaGo and AlphaZero developed by DeepMind, demonstrate remarkable strategic thinking and learning capabilities that showcase their vicinity to general intelligence. These systems excel in creating games, for example, chess, Go, and video games, showcasing adaptive learning and decision-making abilities. While these AI systems demonstrate certain indications of AGI capabilities, such as understanding and problem-solving skills, it is crucial to emphasize that they have not yet achieved AGI status themselves.

7.8 ROBOTICS AND EMBODIED INTELLIGENCE

Robotics and embodied intelligence in the context of AGI involve the integration of AI algorithms with physical robots to enable machines to perceive, interact with, and learn from the physical environment. This integration emphasizes the importance of sensory inputs and motor skills in shaping an understanding of the world through AI systems. The concept of embodied intelligence in robotics proposes that intelligence is not solely a function of algorithms but also the physical manifestation of an entity and its interactions with the environment. By integrating AI with robots, developers aim to create systems that learn from and adapt to the physical world, mirroring how humans and animals learn through interaction and experience. For instance, robots equipped with sensors such as cameras, lidar, radar, or tactile sensors gather data from the environment, providing information about surroundings, objects, and potential obstacles. AI algorithms process this sensory input to make sense of the environment, enabling robots to perceive and understand their surroundings.

Also, robots need the ability to act upon their environment through movement and manipulation. Advanced motor skills involve grasping objects, navigating environments, and performing complex actions. A robot's movements are controlled by AI algorithms, enabling it to interact with and manipulate objects based on its sensory perceptions. The combination of perception and action forms a feedback loop that facilitates learning. As robots interact with the environment, they receive feedback based on their actions, which helps refine their understanding and decision-making processes. Through reinforcement learning, the robots can learn from trial and error, adjusting their behaviors based on the outcomes of their actions in the physical world.

The integration of AI with robots has numerous real-world applications. In manufacturing, AI-powered robots can adapt to changing environments and tasks to optimize production processes. Additionally, robotic systems can assist surgeons in healthcare, aid in rehabilitation, or support individuals with disabilities. Moreover, Boston Dynamics exemplifies embodied intelligence through robots like Spot and Atlas. Spot is an agile robot dog that utilizes sensors and AI to navigate complex terrain and learn from interactions, adapting its movements for improved performance. Whereas, Atlas is an acrobatic humanoid that showcases advanced balance and dexterity, performing complex maneuvers with stability and agility.

However, challenges persist in achieving robust embodied intelligence in robotics, such as developing AI systems that can adapt to diverse and unpredictable real-world scenarios, handle uncertainties, and learn effectively from physical interactions. Additionally, ensuring the safety, reliability, and ethical implications of AI-powered robots operating in real-world settings is critical in this field. Therefore, the synergy between AI and robotics in achieving embodied intelligence represents a significant step toward AGI.

7.9 ARTIFICIAL SUPER INTELLIGENCE

ASI is the hypothetical future of AGI which is expected to possess cognitive abilities far beyond human capacity. This will enable it to solve complex problems, acquire knowledge across multiple domains, and exhibit creativity and consciousness, fundamentally altering the dynamics of society and technology. The theoretical concept of ASI remains speculative, as achieving it poses profound scientific and ethical considerations due to its potential for immense impact on humanity. Notably, concerns arise regarding control over AI systems with such capabilities, posing technological challenges. Therefore, human control over ASI is crucial to prevent unintended consequences and uphold ethical principles. Furthermore, it is essential to address concerns regarding safety, transparency, and ethical alignment in the development and deployment of ASI to mitigate potential risks and promote beneficial outcomes for society.

7.10 SUMMARY

This chapter provided a detailed overview of AGI, beginning with the definition and differentiation from other categories of AI (i.e., ANI and ASI). It then presented the characteristics of AGI, exploring the cognitive functions necessary for an AI system to demonstrate broad intelligence. Various approaches for developing AGI, including symbolic reasoning, artificial neural networks, and hybrids, were presented alongside discussions on the philosophical foundations of AGI in the philosophy of mind. Additionally, the chapter examined the inherent challenges in AGI development and how to address ethical, safety, and control concerns while weighing the potential benefits and risks across different domains. It further scrutinized indicators of AGI presence, such as LLMs, humanoid robots, self-driving cars, and game-playing AI systems, which demonstrate significant advancements in understanding and problem-solving abilities. The chapter also discussed the role of robotics and embodied intelligence in enabling AGI to perceive, interact with, and learn from their environment. Finally, the concept of ASI was briefly explored, envisioning a future where AI surpasses human intelligence significantly, accompanied by a discussion on the associated implications and the imperative for responsible development and governance.

Exercises

1. Describe three key characteristics differentiating general intelligence from narrow or specialized intelligence. Provide examples to illustrate these differences.
2. Choose any two challenges associated with AGI development and propose potential strategies to overcome or mitigate them.
3. Apart from the examples this chapter outlines, describe a recent real-world application or advancement toward AGI.
4. Describe the applications and potential impacts of AGI on society.
5. Discuss potential advancements, challenges, and the implications of AGI over the next decade on various aspects of society, including ethics, employment, and technology.
6. Imagine you could design your own AGI system. What key features and abilities would you prioritize?
7. Describe current AI systems which, to some extent, exhibit early manifestations of AGI.
8. Analyze and compare two philosophical theories or perspectives regarding the nature of consciousness and its emulation in AGI systems.
9. Discuss a recent advancement in robotics technology that showcases embodied intelligence principles.
10. Discuss the potential biases in AGI systems trained on real-world data, how they arise, and propose strategies to mitigate their impact on decision-making and social interactions.

FURTHER READING

Baum, S. (2017). A survey of artificial general intelligence projects for ethics, risk, and policy. Global Catastrophic Risk Institute Working Paper, 17–1.
Bubeck, S., Chandrasekaran, V., Eldan, R., Gehrke, J., Horvitz, E., Kamar, E., Lee, P. et al. (2023). Sparks of artificial general intelligence: Early experiments with GPT-4. arXiv preprint arXiv:2303.12712.
Goertzel, B. (2007). *Artificial general intelligence*. Edited by Cassio Pennachin (Vol. 2). Springer.
Goertzel, B. (2014). Artificial general intelligence: Concept, state of the art, and future prospects. *Journal of Artificial General Intelligence*, 5(1), 1–46.

Goertzel, B., Pennachin, C., & Geisweiller, N. (2014). Engineering general intelligence, Part 1. *Atlantis Thinking Machines*, 5, 1–318.

Goertzel, B., Pennachin, C., & Geisweiller, N. (2014). *Engineering general intelligence, Part 2: The CogPrime architecture for integrative, embodied AGI* (Vol. 6). Springer.

Pei, Jing, Lei, Deng, Sen, Song, Mingguo, Zhao, Youhui, Zhang, Shuang, Wu, & Guanrui, Wang. (2019). Towards artificial general intelligence with hybrid Tianjic chip architecture. *Nature*, 572(7767), 106–111.

Pennachin, C., & Goertzel, B. (2007). Contemporary approaches to artificial general intelligence. In Goertzel, B., Pennachin, C. (eds) *Artificial general intelligence* (pp. 1–30). Springer. https://doi.org/10.1007/978-3-540-68677-4_1

Voss, P. (2007). Essentials of General Intelligence: The Direct Path to Artificial General Intelligence. In Goertzel, B., Pennachin, C. (eds) *Artificial general intelligence*, 131–157. Springer. https://doi.org/10.1007/978-3-540-68677-4_4

Wang, Pei, & Goertzel, B. (Eds.). (2012). *Theoretical foundations of artificial general intelligence* (Vol. 4). Springer Science & Business Media.

Machine learning step-by-step practical examples

8

Upon completing this chapter, learners should be able to:
1. Understand how to approach various machine learning problems.
2. Apply practical data preprocessing skills to address machine learning problems.
3. Apply classification algorithms to classify data into distinct categories and interpret the results.
4. Utilize regression algorithms on real-world datasets to make predictions and evaluate model performance.
5. Apply clustering algorithms to partition real-world data into groups based on similarity and interpret and visualize results.
6. Apply association rules techniques to discover relationships between items in a real-world dataset.

8.1 CASE STUDY 1: CLASSIFICATION PROBLEM

This case study focuses on detecting diabetes using a machine learning classifier, where the data samples are classified into two classes (i.e., positive or negative). The subsequent subsections outline the steps involved in handling this particular case study.

8.1.1 Problem definition

Diabetes is a chronic disease that leads to elevated levels of blood sugar. When this condition develops, individuals may experience a range of uncomfortable, dangerous, and potentially life-threatening symptoms. These symptoms include high blood pressure, increased susceptibility to infections, heart disease risks, gastroparesis, blood vessel damage, malfunctioning of the pancreas, and irreversible blindness.

8.1.1.1 Description of the dataset

The case study utilizes the widely known Pima Indian Diabetes Dataset, a popular dataset for machine learning tasks. This dataset can be used to train, test, and evaluate new machine learning algorithms and develop models for diabetes prediction. It is publicly available for download from the Kaggle data science repository (https://www.kaggle.com/datasets/uciml/pima-indians-diabetes-database). The dataset consists of 768 records of women at least 21 years old. Each record contains nine features (8 input and 1 output/outcome/target) as follows:

- *Pregnancies*: Number of times pregnant
- *Glucose*: Plasma glucose concentration 2 hours in an oral glucose tolerance test
- *BloodPressure*: Diastolic blood pressure (mm Hg)
- *SkinThickness*: Triceps skin fold thickness (mm)
- *Insulin*: Serum insulin concentration (mu U/ml)
- *BMI*: Body mass index (weight in kg/height in m^2)
- *DiabetesPedigreeFunction*: Diabetes pedigree function
- *Age*: Age (years)
- *Outcome*: The target feature is a binary variable (i.e., 1 or 0) indicating whether the patient has diabetes or not (i.e., positive or negative).

8.1.2 Loading libraries

Loading the required libraries for data manipulation and model development is essential. The *import* statement is used to load a library in Python. Therefore, the following code snippet loads the necessary libraries required in this case study. More details about each imported library are provided using comments indicated by the hash sign (#).

```
# importing the pandas library for data manipulation
import pandas as pd

# importing the numpy library for mathematical computations
import numpy as np

# importing the scipy library for data transformation
from scipy.stats import zscore
```

```
# importing the seaborn library for data visualization
import seaborn as sns

# importing the matplotlib library for data visualization
import matplotlib.pyplot as plt

# importing train_test_split function from sklearn library
# for splitting the dataset into the train and test sets
from sklearn.model_selection import train_test_split

# importing logistic regression algorithm from sklearn library
from sklearn.linear_model import LogisticRegression

# importing evaluation metrics from sklearn library
from sklearn.metrics import precision_score, recall_score,
f1_score,accuracy_score, confusion_matrix, ConfusionMatrixDisplay
```

8.1.3 Loading dataset

Once the necessary libraries have been imported, the subsequent step involves loading the dataset file (in this case, *diabetes.csv*) using the *read_csv* method in the panda's library. The dataset should be loaded from its stored file path, which may vary depending on the file's location within the computer being used. For simplicity, storing the dataset file and the code or notebook file in the same directory is advised where there is no need to specify the absolute file path, as seen in the following code snippet. This code snippet shows the content of the first five records (depicted in Figure 8.1) using the *data.head(5)* command statement.

```
# Loading the diabetes dataset
diabetes_data = pd.read_csv("diabetes.csv")

# Displaying the first few records
print("First 5 records:")
print(diabetes_data.head(5))
```

First 5 records:

	pregnant	glucose	bp	skin	insulin	bmi	pedigree	age	Outcome
0	6	148	72	35	0	33.6	0.627	50.0	1
1	1	85	66	29	0	26.6	0.351	31.0	0
2	8	183	64	0	0	23.3	0.672	32.0	1
3	1	89	66	23	94	28.1	0.167	21.0	0
4	0	137	40	35	168	43.1	2.288	33.0	1

FIGURE 8.1 The first five records of the dataset.

```
<class 'pandas.core.frame.DataFrame'>
RangeIndex: 768 entries, 0 to 767
Data columns (total 9 columns):
 #   Column    Non-Null Count  Dtype
---  ------    --------------  -----
 0   pregnant  768 non-null    int64
 1   glucose   768 non-null    int64
 2   bp        768 non-null    int64
 3   skin      768 non-null    int64
 4   insulin   768 non-null    int64
 5   bmi       768 non-null    float64
 6   pedigree  768 non-null    float64
 7   age       765 non-null    float64
 8   Outcome   768 non-null    int64
dtypes: float64(3), int64(6)
memory usage: 54.1 KB
```

FIGURE 8.2 Data summary.

8.1.4 Data summary

After loading the dataset, it is essential to get a summary of the loaded dataset. As shown in the following code snippet, the *info()* method can be used to provide details such as the number of rows and columns, the data types of the columns, and the memory usage of the dataset. Figure 8.2 shows the output of the *data.info()* command statement.

```
diabetes_data.info()
```

8.1.4.1 Descriptive statistics

There are various ways of summarizing and describing the main properties of attributes of the dataset in Python, such as the central tendency, dispersion, and shape. For instance, the *describe()* method is used to display the measures of central tendency for all numerical attributes in the dataset. In this case, the following code snippet is used for such purposes. Figure 8.3 shows the output of the *data.describe()* command statement.

```
diabetes_data.describe().round(2)
```

Additionally, the following code snippet computes and displays the number of records in each class and their corresponding percentages. As it is shown in Figure 8.4, the classes labeled as 0 (i.e., negative) and 1 (i.e., positive) have a total of 500 (65.10%) and 268 (34.90%) records, respectively. These statistics show that the two classes are imbalanced, as the number of records in the negative class is almost double that of the positive class. This gives insightful information to help you understand the class composition of the dataset and consider potential implications for data analysis and modeling.

	pregnant	glucose	bp	skin	insulin	bmi	pedigree	age	Outcome
count	768.00	768.00	768.00	768.00	768.00	768.00	768.00	765.00	768.00
mean	3.85	120.89	69.11	20.54	79.80	31.99	0.47	33.26	0.35
std	3.37	31.97	19.36	15.95	115.24	7.88	0.33	11.76	0.48
min	0.00	0.00	0.00	0.00	0.00	0.00	0.08	21.00	0.00
25%	1.00	99.00	62.00	0.00	0.00	27.30	0.24	24.00	0.00
50%	3.00	117.00	72.00	23.00	30.50	32.00	0.37	29.00	0.00
75%	6.00	140.25	80.00	32.00	127.25	36.60	0.63	41.00	1.00
max	17.00	199.00	122.00	99.00	846.00	67.10	2.42	81.00	1.00

FIGURE 8.3 Descriptive statistics for each column in the dataset.

```
Class Counts:
Outcome
0    500
1    268
Name: count, dtype: int64

Class Percentages:
Outcome
0    65.104167
1    34.895833
Name: count, dtype: float64
```

FIGURE 8.4 The class distribution of the dataset.

```
# Counting the number of samples in each target class
target_counts = diabetes_data['Outcome'].value_counts()
print("\nClass Counts:")
print(target_counts)

# Calculating the percentage of samples in each target class
target_percentages = (target_counts / len(diabetes_data)) * 100
print("\nClass Percentages:")
print(target_percentages)
```

8.1.4.2 Data visualization

It is essential to visually analyze the characteristics of the dataset in order to get insights such as the relationships and comparisons between features, checking for the presence of outliers and other data-relevant insights. The following code snippet displays the visual representation of the class distribution in the dataset. The resultant output is depicted in Figure 8.5.

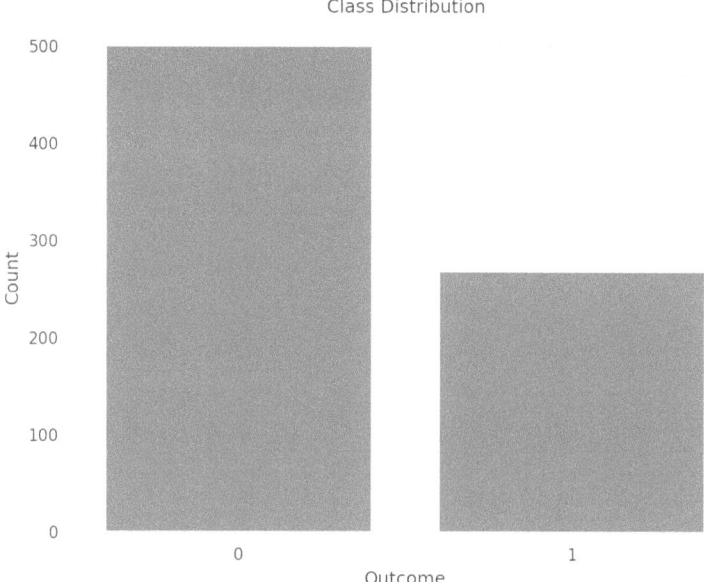

FIGURE 8.5 Class distribution.

```
# Visualizing the class distribution
sns.set(style="whitegrid")
plt.figure(figsize=(8, 6))
plt.title("Class Distribution")
sns.set_palette("Set2")
sns.countplot(x='Outcome', data=diabetes_data)
plt.xlabel("Target Class")
plt.ylabel("Count")
plt.show()
```

Furthermore, scatter plots are commonly used to visualize insights about a dataset, such as correlations between features, outlier detection, and feature distribution. For example, the subsequent code generates a scatter plot that helps examine the relationship between BMI and age features using the *scatterplot()* function. As depicted in Figure 8.6, the resulting scatter plot enabled the discovery of outliers in the dataset, as seen in the two red ovals.

```
plt.figure(figsize=(8, 6))
sns.scatterplot(x="bmi", y="age", data=diabetes_data, sizes=(1, 8),
hue="Outcome")
plt.title("Age against BMI Scatterplot")
plt.show()
```

Moreover, a boxplot visually displays the lower fence, the first quartile (25th percentile), the median (50th percentile), the third quartile (75th percentile), and the upper fence values of the feature, along with any outliers. The following code snippet generates the boxplot showing the BMI feature's outliers. The resulting boxplot, depicted in Figure 8.7, suggests that the BMI values below 18 and above 50 are

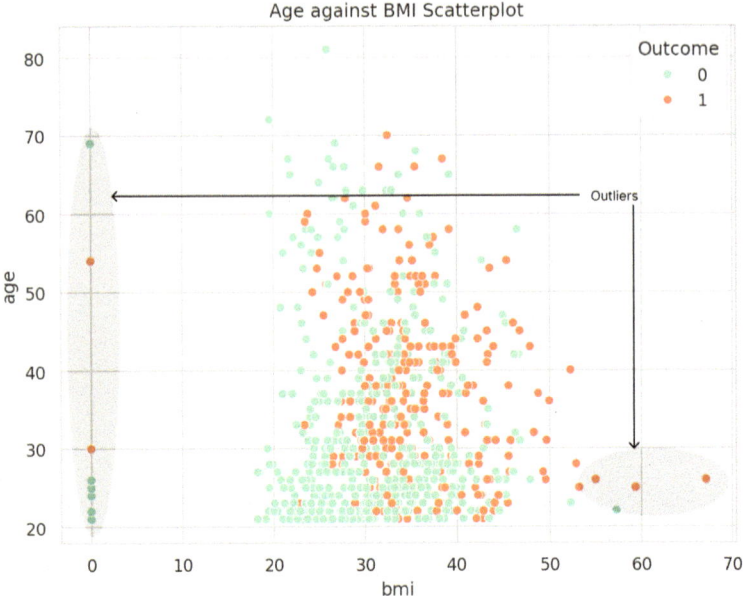

FIGURE 8.6 Scatter plot for age against BMI.

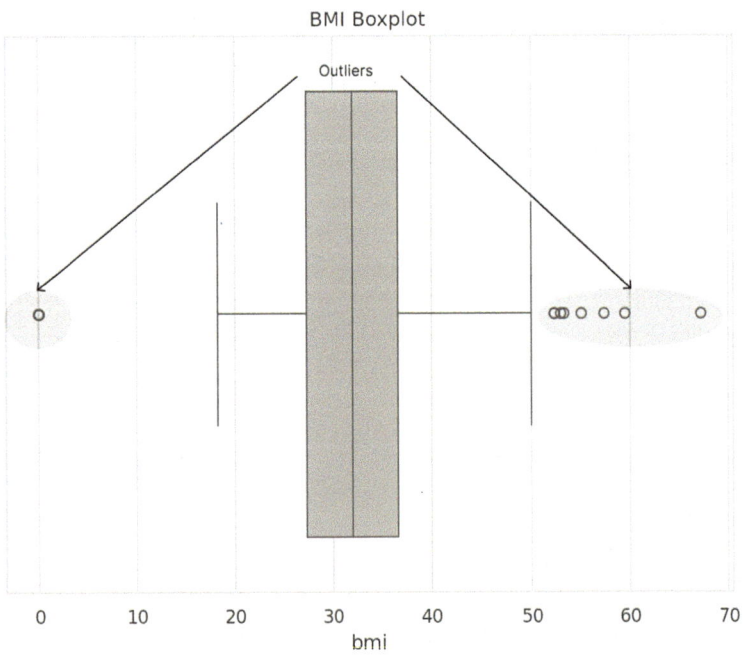

FIGURE 8.7 BMI box plot (with outliers).

considered outliers. It is worth noting that the code snippet for the box plot and scatter plot can also be used to detect the presence of outliers in other dataset features apart from the BMI feature.

```
plt.figure(figsize=(8, 6))
sns.boxplot(x=diabetes_data['bmi'])
plt.title("BMI Boxplot")
plt.show()
```

8.1.5 Data preprocessing

Data preprocessing is vital to cleaning, refining, transforming, and formatting data to ensure its suitability for machine learning tasks. Data preprocessing can significantly impact the effectiveness and accuracy of the developed models, as the quality of the data used directly influences them. As part of data preprocessing, we showcase how to handle outliers and missing values and standardize data to prepare it for subsequent steps.

8.1.5.1 Data cleaning

This section focuses on handling outliers and missing values in the dataset, as detailed below.

8.1.5.1.1 Outliers
The BMI feature in the dataset contains outliers; therefore, outliers for the BMI feature that fall below the lower fence are trimmed because their values are zero. Note that the values within the lower and upper fences can either be trimmed or winsorized (replacing an outlier value with the nearest non-outlier value). However, in this case, the BMI values above the upper fence are replaced with the value of the nearby upper fence. The following code snippet demonstrates removing and trimming outliers and plotting the resultant box plot, as shown in Figure 8.8.

```
# Removing records with bmi value of zero
diabetes_data = diabetes_data.drop(diabetes_data[diabetes_data
['bmi'] == 0].index, axis=0)

# Winsorizing bmi outliers above the upper fence
bmi_upper_fence = 50
diabetes_data['bmi'] = diabetes_data['bmi'].clip(upper=bmi_
upper_fence)

# Visualizing bmi distribution after handling outliers
plt.figure(figsize=(8, 6))
sns.boxplot(x=diabetes_data['bmi'])
plt.title("bmi Distribution (Outliers Removed)")
plt.show()
```

8.1.5.1.2 Missing values
As the dataset for this case study contains no missing values, some values of the "Age" feature are intentionally set to null in the original dataset to demonstrate how to deal

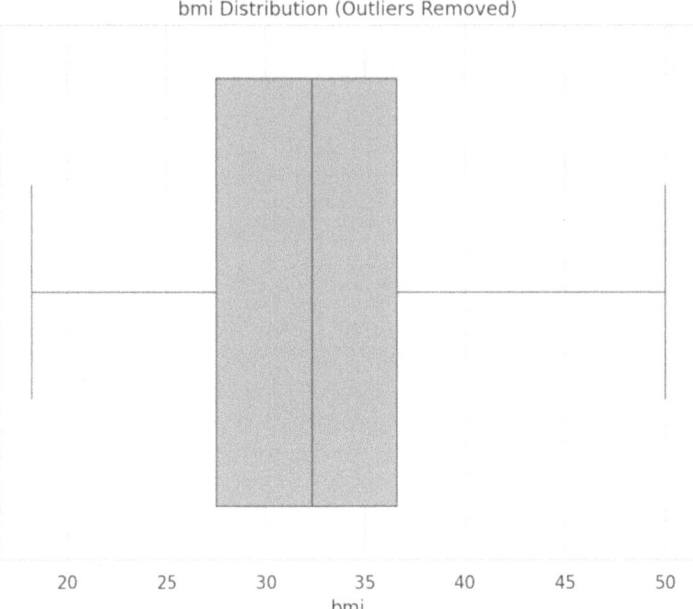

FIGURE 8.8 BMI box plot after removing and trimming outliers.

```
pregnant    0
glucose     0
bp          0
skin        0
insulin     0
bmi         0
pedigree    0
age         3
Outcome     0
dtype: int64
```

FIGURE 8.9 Checking the presence of missing values for all features.

with missing values. Consequently, the resulting dataset contains some missing values. The following code snippet is used to check the presence of missing values for all features in the dataset. Figure 8.9 shows no missing values for all features except for the "Age" feature, which comprises three null values.

```
# Checking for missing values
print(diabetes_data.isnull().sum())
```

Moreover, several methods can be used to handle the identified missing values in the "Age" feature, including imputation by using measures of central tendencies (mean, median, or mode) or removing the corresponding records that contain missing values. In this case, the missing values are filled by the median value of the "Age" feature, as shown in the following code snippet. The median was chosen after examining the mean, mode, and median of the "Age" and determining that the median value was the most suitable for this dataset.

```
# Imputing missing values in the age feature
diabetes_data['age']=diabetes_data['age'].fillna(diabetes_
data['age'].median())
```

8.1.5.2 Data standardization

Before data standardization, the target feature (i.e., Outcome) should be separated from the rest (i.e., input features) as shown in the following code snippet. It is worth noting that the target feature is separated to avoid standardizing its values.

```
# separating the target feature from the input features
predictor_vars = diabetes_data.drop("Outcome", axis=1)
target_var = diabetes_data["Outcome"]
```

Moreover, after separating the target feature, the input features are standardized using a *z-score*, as showcased in the following code snippet. Figure 8.10 depicts the standardized values of the input features. Note the difference between the standardized values (Figure 8.10) and non-standardized values (Figure 8.1).

```
# Standardizing the features
standardized_predictors = predictor_vars.apply(zscore)
# display the first few records
print(standardized_predictors.head())
```

8.1.6 Split-out the dataset

After data preprocessing, the dataset should be split into two sets: a training and a test set. The training set is used to train the model, and the test set is used to evaluate the model's performance. The following code snippet splits the dataset into two sets in a ratio of 80:20 for the training and testing sets (i.e., *test_size=0.2*), respectively. It is

	pregnant	glucose	bp	skin	insulin	bmi	pedigree	age
0	0.640157	0.838444	0.126380	0.894820	-0.699113	0.180170	0.469430	1.432300
1	-0.844521	-1.127834	-0.202005	0.517415	-0.699113	-0.863925	-0.368823	-0.195705
2	1.234029	1.930820	-0.311467	-1.306707	-0.699113	-1.356141	0.606102	-0.110020
3	-0.844521	-1.002991	-0.202005	0.140010	0.113794	-0.640190	-0.927658	-1.052549
4	-1.141457	0.495125	-1.625006	0.894820	0.753742	1.597156	5.514134	-0.024336

FIGURE 8.10 Standardized input features.

worth noting that *random_state=42* sets a value to ensure that the random splitting of the dataset will be reproducible.

```
# Splitting the data into train and test sets
train_predictors, test_predictors, train_targets, test_targets =
train_test_split(standardized_predictors, target_var,
test_size=0.20, random_state=42)
```

8.1.7 Choosing classification algorithm

Notably, there are many classification algorithms; therefore, one needs to spot-check and select just one or a few algorithms that can properly address the problem. Spot-checking explores which algorithm(s) is the best performing on the respective problem. Some popular classification algorithms include Support Vector Machine, Decision Tree, K-Nearest Neighbor (KNN), Logistic Regression, Random Forest, and Naive Bayes. In this case study, the Logistic Regression algorithm was selected due to its simplicity, interpretability, and computational efficiency in modeling the probability of a binary outcome. This algorithm is arbitrarily selected for demonstration purposes.

8.1.8 Training the model

The logistic regression algorithm is trained using the training set, which allows it to learn the relationship between the input and the target features. Therefore, the following code snippet demonstrates the training of the Logistic Regression algorithm.

```
# initialize the instance of the algorithm
logistic_model = LogisticRegression()
# using the instance to train the algorithm
logistic_model.fit(train_predictors, train_targets)
```

8.1.8.1 Model evaluation

It is essential to evaluate model performance on the test set based on different metrics such as Confusion Matrix, Accuracy, Precision, Recall, F-score, Sensitivity, Specificity, ROC, and AUC. The following code snippet evaluates the model performance based on Accuracy, Precision, Recall, and F1-score, and the performance evaluation results are depicted in Figure 8.11.

```
Accuracy: 0.756578947368421
Precision: 0.6875
Recall: 0.6
F1 Score: 0.6407766990291262
```

FIGURE 8.11 Model performance evaluation results.

```
# Making predictions on the test set
test_predictions = logistic_model.predict(test_predictors)

# Computing and printing the performance metrics
print("Accuracy:", accuracy_score(test_targets,
test_predictions))
print("Precision:", precision_score(test_targets,
test_predictions))
print("Recall:", recall_score(test_targets, test_predictions))
print("F1 Score:", f1_score(test_targets, test_predictions))
```

Furthermore, the confusion matrix is also used to evaluate the model's performance by observing the number of predicted labels against the actual labels in a given class. The following code snippet generates the confusion matrix of the model, and the results are depicted in Figure 8.12. Notably, the number of true negatives is 82, false negatives are 15, false positives are 22, and true positives are 33. These results imply that the model can correctly predict many instances of the negative class compared to the positive class.

```
# Visualizing the confusion matrix
cm = confusion_matrix(test_targets, test_predictions,
labels=logistic_model.classes_)
disp = ConfusionMatrixDisplay(confusion_matrix=cm, display_
labels=logistic_model.classes_)
plt.figure(figsize=(8, 6))
disp.plot()
plt.title("Confusion Matrix")
plt.grid(False)
plt.show()
```

8.1.8.2 Saving the model

In machine learning, saving the model involves storing a trained model on a computer storage or external drives, which enables the model to be reused to make predictions

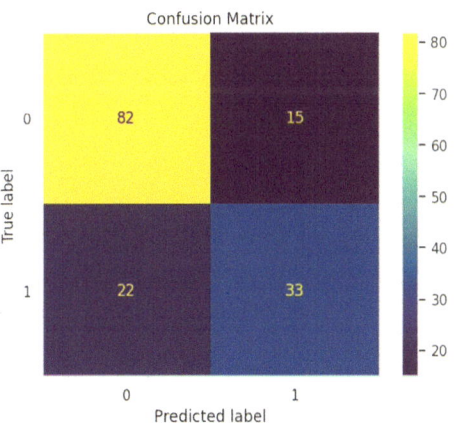

FIGURE 8.12 Confusion matrix.

on new, unseen data without retraining it from scratch. Various libraries, such as joblib and pickle, can be used to save the trained model. The following code snippet demonstrates how the joblib library using the *dump()* method is employed to save the trained model in the current working directory with the file named *logistic_model.joblib*.

```
# Saving the trained model
joblib.dump(logistic_model, 'diabetes_logistic_model.joblib')
```

8.1.8.3 Inferencing

Once a model has been trained and saved, it can be used to classify/predict new, unseen data that were not part of the training and test sets. During predictions, the new data has to undergo the same data preprocessing steps applied during the training phase. Suppose the unseen data needs to be classified using the saved model; it will have to be checked for outliers and standardized using the *zscore()* function as presented in the previous steps. As shown in the following code snippet, the new, unseen data has been classified as 1 (i.e., positive) after undergoing the necessary preprocessing steps and being fed into the trained and saved model. Figure 8.13 shows that the new, unseen data has been classified as 1 (i.e., positive).

```
# load the model
loaded_model = joblib.load('diabetes_logistic_model.joblib')

# Calculate the mean and standard deviation of each feature from
the training data
feature_means = diabetes_data.drop("Outcome", axis=1).mean()
feature_stds = diabetes_data.drop("Outcome", axis=1).std()

# Defining new, unseen data
new_data = [6, 148, 72, 35, 0, 33.6, 0.627, 50]  # Example
realistic data

# Standardize the new data using the means and standard
deviations from the training data
standardized_new_data = (new_data - feature_means) / feature_stds

# Reshaping the standardized new data
reshaped_new_data = standardized_new_data.values.reshape(1, -1)

# Creating a DataFrame with feature names and standardized new data
feature_names = ['pregnant', 'glucose', 'bp', 'skin', 'insulin',
'bmi', 'pedigree', 'age']
new_data_df = pd.DataFrame(reshaped_new_data,
columns=feature_names)

# Making predictions on the standardized new data
print("\nPrediction on New Data:")
print("The new data is predicted as class : ", loaded_model.
predict(new_data_df)[0])
```

```
Prediction on New Data:
The new data is predicted as class :  1
```

FIGURE 8.13 Result of new data prediction.

8.2 CASE STUDY 2: REGRESSION PROBLEM

This case study focuses on regression analysis using an advertising dataset. This problem demonstrates the relationship between advertising and sales and aims to develop a model to predict sales based on advertising budgets. The following subsections outline the steps in developing a prediction model using this dataset.

8.2.1 Problem definition

Sales prediction through advertising on TV, radio, and a newspaper is complex due to a number of factors that can influence sales, including the target audience, message, medium, budget, and the timing of the advertising campaign. Consequently, it becomes challenging to accurately predict how much sales will increase as a direct outcome of advertising. This section aims to show step-by-step how to develop a regression model that can predict sales based on advertising on TV, radio, and a newspaper.

8.2.1.1 Description of the dataset

The advertising dataset used in this case study is a collection of structured data that contains information related to advertising costs across multiple channels, including radio, TV, and newspapers. The dataset is used to understand the correlation between advertising expenditures and the generated sales revenue. It also compares the effectiveness of different advertising channels (i.e., TV, radio, and newspaper). The dataset is publicly available for download from the Kaggle data science repository (https://www.kaggle.com/datasets/tawfikelmetwally/advertising-dataset). The dataset contains 200 rows and the following four features:

- *TV*: The amount spent on TV advertisements.
- *Radio*: The amount spent on radio advertisements.
- *Newspaper*: The amount spent on newspaper advertisements.
- *Sales*: The target feature shows the total sales revenue generated.

8.2.2 Loading libraries

As pointed out in Case Study 1, importing the required libraries for data manipulation and model development is essential. The following code snippet imports the required libraries in this case study. Again, more details about each library are provided using comments indicated by the hash sign (#).

```
# library to store data
import pandas as pd
```

```
# library to perform mathematical #computations on matrices
import numpy as np

# library to calculate #standardization
from scipy.stats import zscore

# library to visualize data
import seaborn as sns

# library to visualize data
import matplotlib.pyplot as plt

# library #to split the data in train and test data
from sklearn.model_selection import train_test_split

# library #to use for machine learning (eg., here logistic
regression) algorithm
from sklearn.linear_model import LinearRegression

# importing the joblib library for model saving
import joblib
```

8.2.3 Loading dataset

After importing the required libraries, the next step is to load the dataset file (i.e., *advertising.csv*) from its stored file path using the *read_csv* function in the pandas library, as shown in the following code snippet. The output of the code snippet is displayed in Figure 8.14, showing three records using the *head()* function.

```
advertising_data = pd.read_csv("advertising.csv")
```
```
# displaying the first three records
advertising_data.head(3)
```

8.2.4 Data summary

The *info()* method is used to display the contents and gain key insights into the dataset. As shown in the following code snippet, the *info()* method displays the number of rows and columns, the data types of the columns, and the memory usage of the dataset. Figure 8.15 shows the output of the *advertising_data.info()* command statement.

```
advertising_data.info()
```

	TV	Radio	Newspaper	Sales
0	230.1	37.8	69.2	22.1
1	44.5	39.3	45.1	10.4
2	17.2	45.9	69.3	9.3

FIGURE 8.14 The first three records of the dataset.

```
<class 'pandas.core.frame.DataFrame'>
RangeIndex: 200 entries, 0 to 199
Data columns (total 4 columns):
 #   Column     Non-Null Count  Dtype
---  ------     --------------  -----
 0   TV         200 non-null    float64
 1   Radio      199 non-null    float64
 2   Newspaper  199 non-null    float64
 3   Sales      200 non-null    float64
dtypes: float64(4)
memory usage: 6.4 KB
```

FIGURE 8.15 Data summary.

8.2.4.1 Descriptive statistics

As demonstrated in Case Study 1, descriptive statistics are used to summarize and describe the main features of a dataset. Therefore, in this case study, the *describe()* method is again used to display the measures of central tendency for all numerical columns in the dataset, as shown in the following code snippet. Figure 8.16 shows the output of the *advertising_data.describe()* command statement.

```
advertising_data.describe()
```

8.2.4.2 Data visualization

Data visualization techniques are used to visually analyze the features of the dataset in order to get insights such as the relationships, comparisons between features and other data-relevant insights. The following code snippet displays the visual representation of the correlation matrix (e.g., showing the relationship among the features). The resultant output is depicted in Figure 8.17. Note that the results in Figure 8.17 show that values closer to 1 indicate stronger positive relationships, while values closer to 0 suggest weaker or no linear relationships.

```
# compute the correlation matrix
corr = advertising_data.corr()

# heatmap with annotations
plt.figure(figsize=(7,7))
plt.title("Correlation among features")
sns.heatmap(corr, annot=True, cmap="coolwarm", fmt=".2f",
square=True, linewidths=.5, cbar_kws={"shrink": .5})
plt.show()
```

In addition, the following code snippet displays the visual representation of the scatter plots to show the relationship between the target and the input features. The resultant output is depicted in Figure 8.18. Note that the steeper the slope of the regression line fitted through the data points in a scatter plot, the stronger the correlation between the features in the dataset, as illustrated in Figure 8.18.

	TV	Radio	Newspaper	Sales
count	200.000000	199.000000	199.000000	200.000000
mean	147.042500	23.370352	30.665829	14.022500
std	85.854236	14.807682	21.775905	5.217457
min	0.700000	0.000000	0.300000	1.600000
25%	74.375000	10.050000	12.850000	10.375000
50%	149.750000	23.300000	25.900000	12.900000
75%	218.825000	36.550000	45.100000	17.400000
max	296.400000	49.600000	114.000000	27.000000

FIGURE 8.16 Descriptive statistics for TV, radio, newspaper, and sales columns in the dataset.

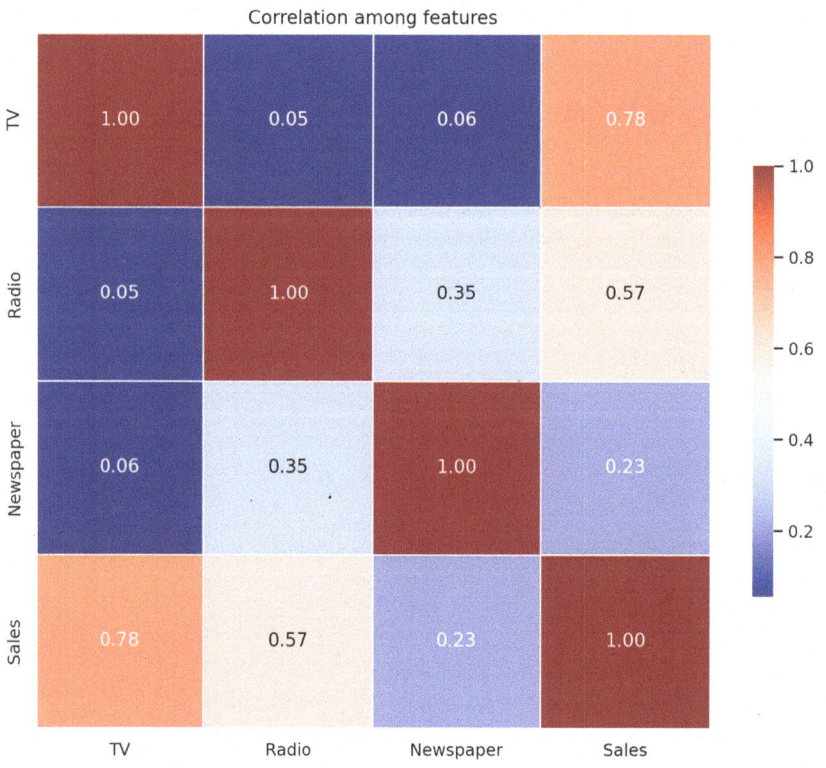

FIGURE 8.17 Correlation matrix.

8 • Machine learning step-by-step practical examples 179

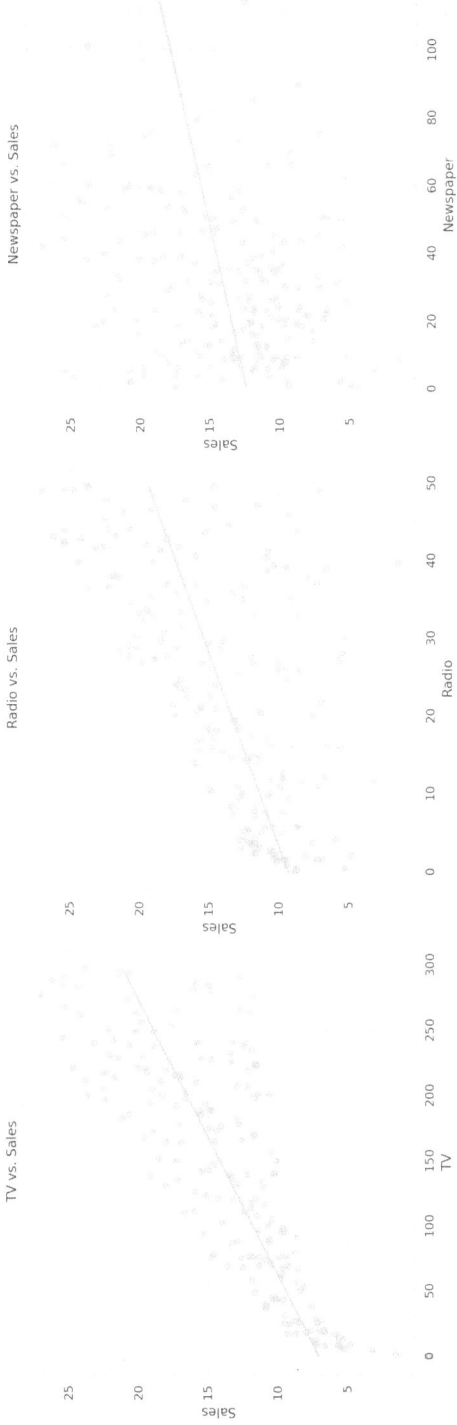

FIGURE 8.18 Scatter plots for the target against the input features.

```
import matplotlib.pyplot as plt
import seaborn as sns

# Create a figure with three subplots
fig, axes = plt.subplots(1, 3, figsize=(18, 6))

# Scatter plot: TV against Sales
sns.regplot(x="TV", y="Sales", data=advertising_data, ax=axes[0])
axes[0].set_title("TV vs. Sales")

# Scatter plot: Radio against Sales
sns.regplot(x="Radio", y="Sales", data=advertising_data,
ax=axes[1])
axes[1].set_title("Radio vs. Sales")

# Scatter plot: Newspaper against Sales
sns.regplot(x="Newspaper", y="Sales", data=advertising_data,
ax=axes[2])
axes[2].set_title("Newspaper vs. Sales")

# Adjust spacing between subplots
plt.tight_layout()

# Display the figure
plt.show()
```

8.2.5 Data preprocessing

In this case study, the data preparation techniques are applied to handle outliers and missing values, along with data transformation, to ensure its readiness for subsequent steps in the modeling phase.

8.2.5.1 Data cleaning

In this case study, the implemented data cleaning methods aim to explore possibilities for handling outliers, addressing missing values, and executing data transformations.

8.2.5.1.1 Outliers
The provided code snippet generates a box plot highlighting outliers, as depicted in Figure 8.19. It is evident from the dataset that only the "newspaper" attribute contains two outlier points. These outliers constitute a small proportion relative to the dataset's overall size and are especially noteworthy in the context of the regression problem at hand. Most regression algorithms exhibit reduced sensitivity to outliers, and since the dataset includes occasionally plausible values, their presence is considered for analysis. It is not a strict rule that outliers must be removed from the dataset on every occasion. The outliers were not removed, imputed, or transformed in this specific use case.

```
advertising_data.plot.box(figsize=(5,5))
```

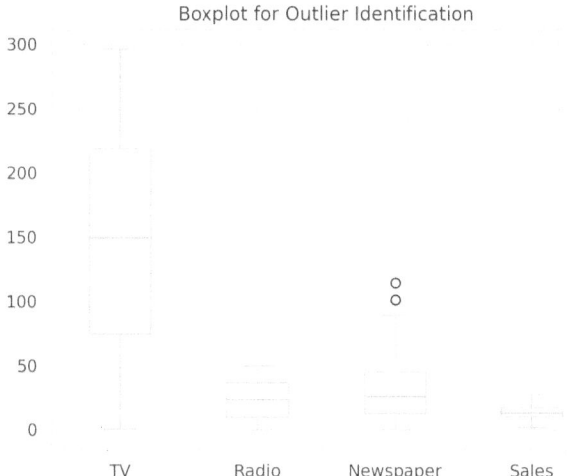

FIGURE 8.19 Box plot for outlier identification.

```
TV          0
Radio       0
Newspaper   0
Sales       0
dtype: int64
```

FIGURE 8.20 Checking the presence of missing values for all features.

8.2.5.1.2 Missing values
The following code snippet is used to assess the presence of missing values across all dataset features to handle missing values. As illustrated in Figure 8.20, no missing values are detected for any of the features. Consequently, no techniques will be applied to handle missing values.

```
advertising_data.isnull().sum()
```

8.2.5.2 Feature selection

The previous correlation matrix in this case study indicates that the features are not highly correlated, as their correlation value is less than 0.35. Based on the correlation values, two features with high correlation values (i.e., above 0.7 or 0.8) might be redundant in providing information to the model. Therefore, one can be eliminated. In this dataset, all the input features can be retained (i.e., none should be eliminated based on the correlation value) since their correlation values are less than 0.35. It is important to note that when there is a zero correlation value between the independent variables and the dependent variable, it necessitates the elimination of the independent variable. The correlation values between sales and TV, Radio, and Newspaper are 0.9, 0.35, and 0.16, respectively (i.e., all are not equal to zero). Therefore, none of the attributes (i.e., TV, Radio and Newspaper) are dropped from the dataset.

8.2.5.3 Data transformation

Upon examining the range of values in the features within the dataset, it becomes apparent that they are not significantly disparate. For example, the "Radio" feature ranges from 0 to approximately 50, "Newspapers" from 0 to about 115, and "Sales" from approximately 1.6 to 27. The only feature that notably stands out is "TV," ranging from 0.7 to nearly 300. Despite the scales not differing drastically (except for "TV"), and considering that the dataset pertains to advertising data where higher values are expected for TV advertisement due to its broader reach, it might be acceptable to forgo normalization. However, it is suggested to experiment with normalization and document any observed differences at the conclusion of the model training.

8.2.6 Choosing regression algorithm

Given the plethora of regression algorithms available, it becomes crucial to spot-check to discern and choose the most suitable algorithm(s) for addressing a specific problem. In this case study, the Multivariate Linear Regression algorithm is selected due to the presence of multiple independent variables and a single dependent variable. It can offer insight into complex relationships within the data.

8.2.7 Training the model

The dataset is divided into training and test sets containing the independent and dependent variables, denoted as X_train, X_test, y_train, and y_test, respectively. Specifically, the dataset is split into a training set comprising 75% of the data and a test set comprising 25% of the data. Such splitting is done using the following code snippet.

```
# X_train and y_train will be used for training the model,
# X_test for testing the models predictions, y_test for
evaluating the model predictions.
X_train, X_test, y_train, y_test = train_test_split(features,
target, test_size=0.25, random_state=42)
```

The model is initialized after splitting the data into training and test sets. Since the Multivariate Linear Regression model is used, the initialized model is *LinearRegression()* and is trained using the training set as shown in the following code snippet.

```
# initialize the instance of the algorithm
lr_model = LinearRegression()

# using the instance to training the algorithm
lr_model.fit(X_train, y_train)
```

```
Intercept: 2.778303460245283
```

FIGURE 8.21 Output showing the *y*-intercept value of the linear regression model.

```
Coefficients: [0.04543356 0.19145654 0.00256809]
```

FIGURE 8.22 Output showing the coefficients of the linear regression model.

Finding the model equation (estimated Sales) starts by finding the estimated regression coefficients. The first regression coefficient is the *y-intercept* and is computed in the following code snippet. In the code snippet, the value of the *y-intercept* is estimated equal to 2.778303460245283, as shown in the subsequent output.

```
lr_model.intercept_
```

The other regression coefficients are computed from the following code snippet. Here, the values of coefficients of TV, Radio, and Newspapers are estimated to equal 0.04543356, 0.19145654, and 0.00256809, respectively, as shown in the subsequent output in Figure 8.22.

```
lr_model.coef_
```

8.2.7.1 Model equation

After estimating the regression coefficients, the equation of the model can now be determined. Using the value of the regression coefficients, the estimated Sales can be computed as follows:

Estimated / Predicted Sales
$$= 2.7783 + 0.0454 * TV + 0.1914 * Radio + 0.0026 * Newspaper$$

8.2.7.2 Evaluating the model

After obtaining the model equation, it is essential to evaluate model performance using different performance metrics for regression problems. These metrics include Mean Square Error (MSE), Root Mean Squared Error (RMSE), Mean Absolute Error (MAE), Coefficient of Determination (R^2 or R-Square), Adjusted R-squared, Mean Percentage Error (MPE), and Coefficient of Variation (CV). For demonstration purposes, only MSE and R-Square are used to evaluate the model.

8.2.7.3 Evaluating the model using MSE

In the following code snippet, the *X_test* represents independent variables used to predict the value of the dependent variable (Sales), here *y_predict*. The predicted dependent variable (*y_predict*) and actual dependent variable (*y_test*) are subjected to the MSE function (`mean_squared_error()`) to calculate the value of MSE. The value

```
                Mean Squared Error (MSE): 2.880023730094193
```

FIGURE 8.23 Output showing the mean squared error (MSE) of the linear regression model.

```
                R-squared: 0.8935163320163657
```

FIGURE 8.24 Output showing the R-squared value of the linear regression model.

of the MSE obtained is 2.880023730094193, as shown in the subsequent output in Figure 8.23. This indicates better model performance in terms of prediction accuracy since it has a lower value.

```
# using the trained model to make 'y_predict' on
# new input features from the test set
y_predict = lr_model.predict(X_test)

# computing and printing the performance metrics
mse = mean_squared_error(y_test, y_predict)
print("Mean Squared Error (MSE):", mse)
```

8.2.7.4 Coefficient of determination

The following code snippet calculates the Coefficient of Determination (R^2) value and the resulting value is 0.8935163320163657, as shown in the subsequent output in Figure 8.24. This means that the model can better explain the variability in the dependent variable.

```
print('R-squared:', r2_score(y_test, predictions))
```

8.3 CASE STUDY 3: CLUSTERING PROBLEM

This case study focuses on the Clustering Problem, which aims to uncover and organize unlabeled data into distinct groups based on inherent similarities or patterns. As previously stated, unlike classification or regression problems where data points already have assigned labels, clustering algorithms must categorize unlabeled data into groups (i.e., clusters). The following subsections outline the steps in developing a clustering model using the given dataset.

8.3.1 Problem definition

The clustering problem in this case study focuses on customer segmentation in malls and shopping complexes. Malls and shopping complexes often compete with each other to increase their customer base in order to increase profit. Segmenting customers proves challenging due to the complex nature of customer behavior, variability in individual preferences, lack of clear understanding of the target audience, and ineffective segmentation criteria. These complexities may lead to datasets with quality issues and potential biases. Achieving effective customer segmentation demands a sophisticated approach covering advanced data handling, robust validation, and domain expertise to navigate these challenges.

8.3.1.1 Description of the dataset

The dataset used in this case study is known as the "Mall Customer Segmentation," a popular choice for developing a model for customer segmentation. It is publicly available for download from the Kaggle data science repository (https://www.kaggle.com/code/listonlt/mall-customers-segmentation-k-means-clustering). The dataset contains five features and 200 samples (i.e., data points) representing individual customers. The features include:

- *CustomerID*: Unique identifier for each customer
- *Gender*: Male or Female
- *Age*: In years (range: 18–70)
- *Annual Income*: In thousands of dollars (range: 15–150)
- *Spending Score*: Reflects customer spending habits (range: 1–100).

8.3.2 Loading libraries

The following code snippet imports the necessary libraries for this specific case study.

```
# Data manipulation libraries
import pandas as pd
import numpy as np

# Data visualization libraries (plotly for interactive graphs)
import matplotlib.pyplot as plt
import seaborn as sns
import plotly.express as px

# Importing sklearn library to use K-Mean algorithm
from sklearn.cluster import KMeans

# Suppresses warnings of type FutureWarning
import warnings
warnings.filterwarnings('ignore', category=FutureWarning)
```

8.3.3 Loading the dataset

Once the essential libraries are imported, the subsequent step involves loading the dataset file (i.e., *mall_customers.csv*) utilizing the '*read_csv*' function within the pandas library. The following code snippet loads the dataset, and Figure 8.25 displays the output, showing five randomly sampled records using the '*customer_data.sample(5)*' function in pandas.

	CustomerID	Gender	Age	Annual Income (k$)	Spending Score (1-100)
63	64	Female	54	47	59
181	182	Female	32	97	86
148	149	Female	34	78	22
150	151	Male	43	78	17
105	106	Female	21	62	42

FIGURE 8.25 Displaying five data samples.

```
customer_data = pd.read_csv("Mall_Customers.csv")
customer_data.sample(5)
```

8.3.4 Renaming column names

The dataset used in this case study contains columns with spaces in their names that need to be renamed for clean and efficient data handling. Spaces in column names can cause issues accessing them using dot notation (*e.g., dataframe.Spending Score*). Additionally, short and descriptive names make the code easier to read and understand, improve the clarity of visualizations, and reduce typo possibilities. The following code snippet renames the column '*Spending Score (1-100)*' to '*Spending_Score*' and the column '*Annual Income (k$)*' to '*Annual_Income*'.

```
customer_data.rename (columns = {
    'Spending Score (1-100)':'Spending_Score',
    'Annual Income (k$)': 'Annual_Income'},
    inplace=True)
```

8.3.5 Data summary

As demonstrated earlier, the '*info()*' method in the following code snippet provides essential data summary details. Figure 8.26 showcases the output of the '*customer_data.info()*' command statement.

```
customer_data.info()
```

8.3.6 Dropping less informative features

In this case study, the *CustomerID* column has to be dropped as it is redundant and non-predictive and does not contribute to understanding the target variable. Eliminating it reduces noise, mitigates overfitting risks, and streamlines computational efficiency during model training and prediction. The '*CustomerID*' in the dataset is dropped using the following code snippet.

```
customer_data.drop("CustomerID", axis=1, inplace=True)
```

```
<class 'pandas.core.frame.DataFrame'>
RangeIndex: 200 entries, 0 to 199
Data columns (total 5 columns):
 #   Column          Non-Null Count  Dtype
---  ------          --------------  -----
 0   CustomerID      200 non-null    int64
 1   Gender          200 non-null    object
 2   Age             200 non-null    int64
 3   Annual_Income   200 non-null    int64
 4   Spending_Score  200 non-null    int64
dtypes: int64(4), object(1)
memory usage: 7.9+ KB
```

FIGURE 8.26 Data summary.

	CustomerID	Age	Annual_Income	Spending_Score
count	200.000000	200.000000	200.000000	200.000000
mean	100.500000	38.850000	60.560000	50.200000
std	57.879185	13.969007	26.264721	25.823522
min	1.000000	18.000000	15.000000	1.000000
25%	50.750000	28.750000	41.500000	34.750000
50%	100.500000	36.000000	61.500000	50.000000
75%	150.250000	49.000000	78.000000	73.000000
max	200.000000	70.000000	137.000000	99.000000

FIGURE 8.27 Descriptive statistics.

8.3.6.1 Descriptive statistics

The following code snippet displays the output of the '*customer_data.describe()*' command statement, as presented in Figure 8.27.

```
customer_data.describe()
```

Additionally, the following code snippet aims to ascertain whether the characteristics of the '*Gender*' feature impact a customer's spending behavior. The '*Gender*' feature is used as it is the only categorical feature in the dataset. The output of the code is displayed in Figure 8.28.

```
# seeks to answer whether gender influences spending
pd.pivot_table(customer_data,index=["Gender"], values=
["Spending_Score"], aggfunc=["count","sum","max","mean"])
```

8.3.6.2 Data visualization

The following code snippet generates a histogram illustrating the relationship between the categorical feature (i.e., '*Gender*') and the total number of samples. The resulting output is presented in Figure 8.29.

```
sns.countplot(x=customer_data["Gender"], data= customer_data)
```

	count	sum	max	mean
	Spending_Score	Spending_Score	Spending_Score	Spending_Score
Gender				
Female	112	5771	99	51.526786
Male	88	4269	97	48.511364

FIGURE 8.28   Relationship between gender spending habit.

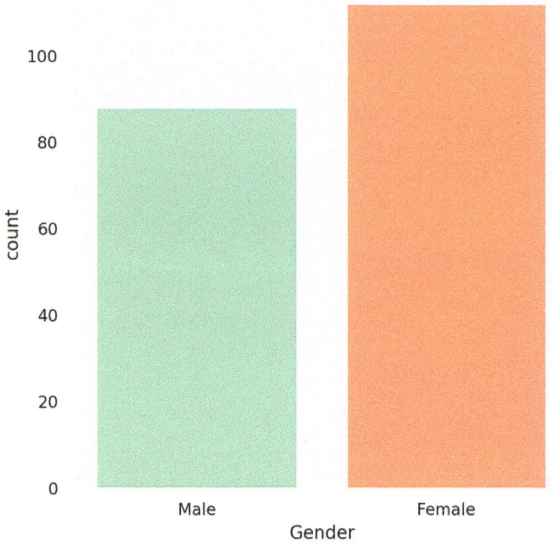

FIGURE 8.29   Relationship between 'Gender' and the total number of samples.

Moreover, the following code snippet creates three subplots, as displayed in Figure 8.30, each containing a distribution plot for each numerical feature in the dataset. This helps in showing the spread of data and detecting outliers by considering deviations from the mean. Note that the data is skewed if the graph leans to one side. The graph's "peakedness" reflects how concentrated the data is around the center. Points far away from the central tendency (mean or median) on the tails of the distribution are potential outliers.

```
# create a single figure with multiple axes to fit the graphs
fig, axs = plt.subplots(1, 3, figsize=(15, 5))
sns.histplot(customer_data["Spending_Score"], kde=True, ax=axs[0])
axs[0].set_title('Spending Score Distribution')
sns.histplot(customer_data["Annual_Income"], kde=True, ax=axs[1])
axs[1].set_title('Annual Income Distribution')
sns.histplot(customer_data["Age"], kde=True, ax=axs[2])
axs[2].set_title('Age Distribution')

# Adjust the padding between and around the subplots
plt.tight_layout()
```

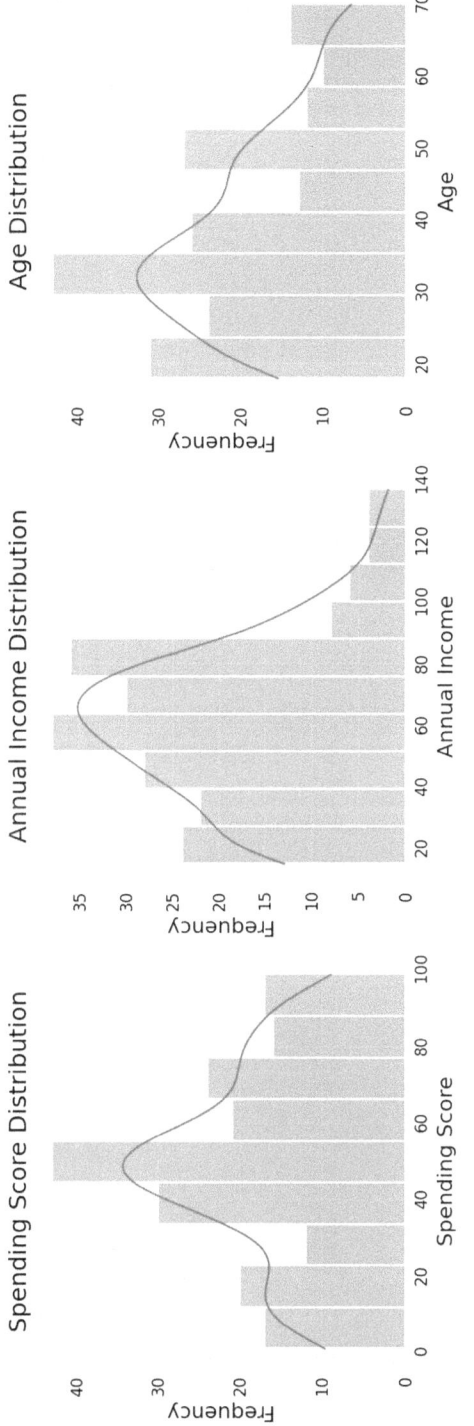

FIGURE 8.30 Distribution plots for numerical features.

Furthermore, the following code snippet generates a scatter plot, illustrating the relationship between the *'Age'* and *'Spending_Score'* features. Observations on the scatterplot in Figure 8.31 suggest a weak correlation between these two features, as represented by data points scattered randomly across the plot without forming a clear pattern or trend. This inference is further illustrated by the correlation matrix depicted in Figure 8.32.

```
plt.figure(figsize=(10,5))
sns.scatterplot(x=customer_data["Age"],y=customer_data
["Spending_Score"])
```

Also, the correlation between the features can further be visualized in the correlation matrix. The following code snippet plots the correlation matrix to visualize the relationships among the numerical features. The matrix provides a more quantified perspective on the relationship between *'Age'* and *'Spending_Score'*, reinforcing the observations made from the scatter plot. For instance, upon examining the correlation value of *'Age'* and *'Spending_Score'*, it can be noted that a correlation value of -0.33 between these two features suggests a slight negative correlation.

```
corr = customer_data.drop("Gender", axis=1).corr()
sns.heatmap(corr, annot=True)
plt.show()
```

8.3.7 Feature transformation

Since the 'Gender' feature is categorical, it needs to be transformed into numerical data before being used to train the model, as most machine learning algorithms work best with numerical data. Therefore, the 'Gender' feature is converted from categories

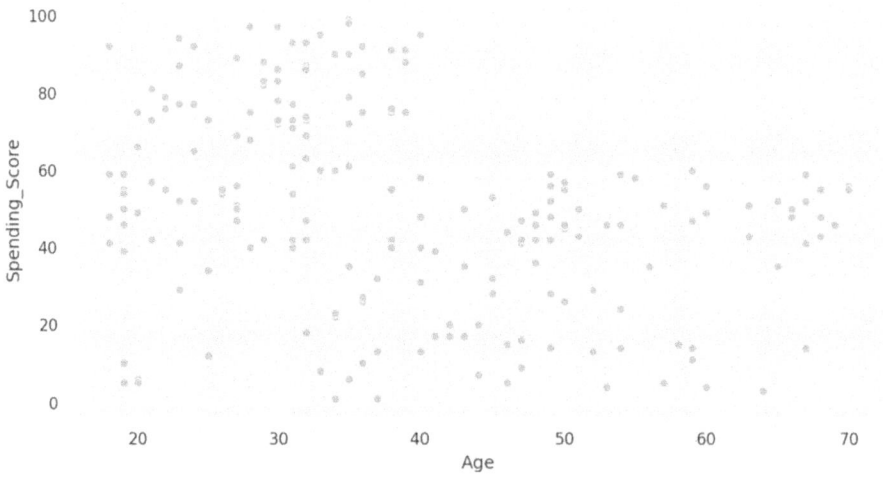

FIGURE 8.31 Distribution plots for numerical features.

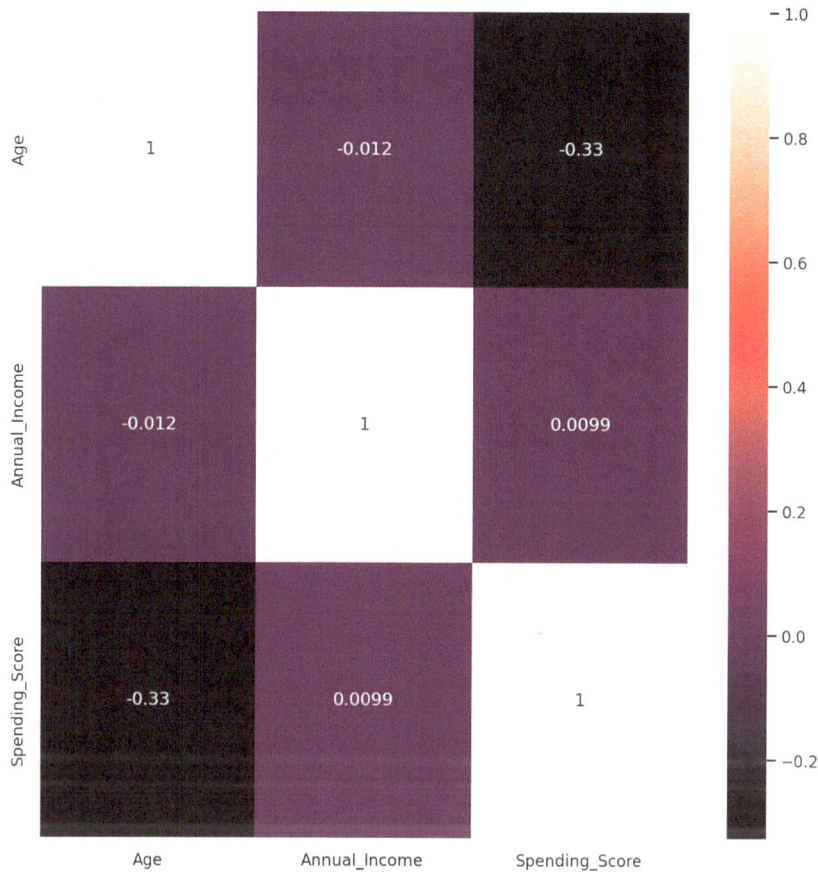

FIGURE 8.32 Correlation matrix.

	Age	Annual Income (k$)	Spending Score (1-100)	Gender_Male
138	19	74	10	1
194	47	120	16	0
172	36	87	10	1
190	34	103	23	0
5	22	17	76	0

FIGURE 8.33 Encoding of the gender feature.

to numbers using one-hot encoding. This technique assigns a unique binary code to each gender (i.e., [1, 0] for 'male' and [0, 1] for 'female'). The following code snippet encodes gender variables using one-hot encoding, and the output is presented in Figure 8.33.

```
# Convert categorical variable(s), in our case Gender, into encoded
# variables, dropping the first category to avoid
multicolinearity
customer_data = pd.get_dummies(customer_data,drop_first=True)
customer_data.sample(4)
```

8.3.8 Performing clustering using K-means algorithm

In this case, the K-means clustering algorithm was chosen for its simplicity. Before conducting the clustering process, the elbow method was utilized to determine the optimal number of clusters (k). The following code snippet illustrates the application of the elbow method to select the most suitable number of clusters. Figure 8.34 showcases the ideal cluster quantity (6 clusters) identified through the elbow method. Note that

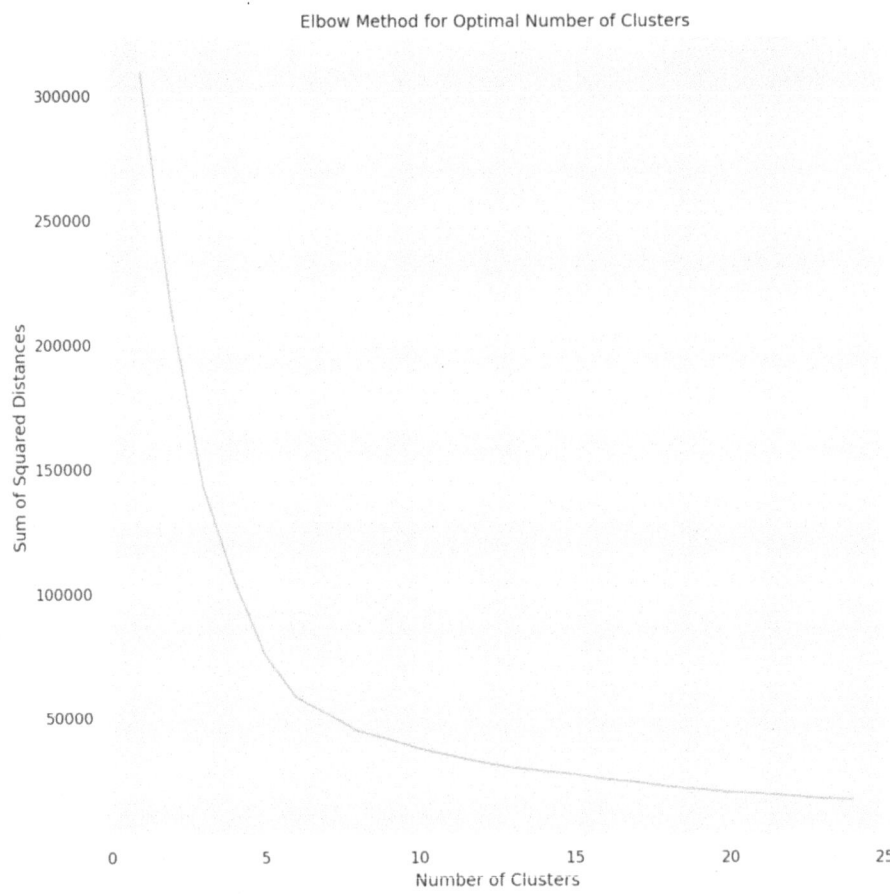

FIGURE 8.34 The optimal number of clusters using the elbow method.

the elbow method identifies the optimal number of clusters at the point where the graph forms an elbow and maintains consistency.

```
cluster_range = range(1, 25)
inertia_values = []
for k in cluster_range:
    cluster_model = KMeans(n_clusters=k)
    cluster_model.fit(customer_data)
    cluster_predictions = cluster_model.predict(customer_data)
    inertia_values.append(cluster_model.inertia_)
plt.plot(cluster_range, inertia_values)
plt.xlabel('Number of Clusters')
plt.ylabel('Sum of Squared Distances')
plt.show()
```

Given the optimal number of clusters generated by the elbow method, the following code snippet performs clustering using the K-means algorithm with the derived optimal number of clusters, which is 6.

```
final_model=KMeans(6)
final_model.fit(customer_data)
prediction=final_model.predict(customer_data)

#Append the prediction
customer_data["GROUP"] = prediction
print("Groups Assigned : \n")
```

The following code snippet renames the group names from numbers to letters for easier readability and visualization. In addition, the code snippet assigns a cluster value to each record in the dataset, simplifying the process of allocating data samples to their respective clusters among the six identified clusters (0 to 5).

```
# Define a mapping from numbers to letters
group_dict = {0: 'A', 1: 'B', 2: 'C', 3: 'D', 4: 'E', 5: 'F'}
# Apply the mapping to the 'GROUP' column
customer_data['GROUP'] = customer_data['GROUP'].map(group_dict)
```

The following code snippet computes the mean of each cluster, as illustrated in Figure 8.35.

```
data_mean = customer_data.drop("Gender_Male", axis=1).groupby
(['GROUP'])
data_mean.mean()
```

The mean values of the identified clusters reveal distinct customer profiles, providing insights that are valuable for tailoring targeted marketing approaches, as described in the following:

- **Group A**, characterized by an average age of 32.69 years, an average annual income of $86.53k, and a high spending score of 82.12, represents middle-aged individuals with high income and spending capacity, suggesting they may be the primary target for luxury goods.

	Age	Annual_Income	Spending_Score
GROUP			
A	27.000000	56.657895	49.131579
B	32.692308	86.538462	82.128205
C	41.685714	88.228571	17.285714
D	25.272727	25.727273	79.363636
E	56.155556	53.377778	49.088889
F	44.142857	25.142857	19.523810

FIGURE 8.35 The mean of each cluster.

- **Group B**, with younger demographics and moderate income and spending tendencies (average age: 27.00 years, average annual income: $56.65k, spending score: 49.13), could be interested in trendy or affordable products.
- **Group C**, comprising older individuals with moderate income and spending scores (average age: 56.16 years, average annual income: $53.38k, spending score: 49.09), may respond well to marketing strategies emphasizing value-oriented products.
- **Group D**, exhibiting middle-aged demographics with high income but lower spending scores (average age: 41.68 years, average annual income: $88.22k, spending score: 17.28), might prefer cautious spending or saving.
- **Group E**, with an average age of 44.14 years, an average annual income of $25.14k, and a spending score of 19.52, consists of older individuals with lower income and spending scores, indicating a preference for discounted or value-oriented products.
- **Group F**, representing younger demographics with lower income but high spending scores (average age: 25.27 years, average annual income: $25.72k, spending score: 79.36), may be inclined toward trendy or impulse purchases.

8.3.9 Cluster visualization

For visualization purposes, the Plotly library displays the data samples in their respective clusters (using the command '*px.scatter()*'). Since k-means uses all data features, visualizing high dimensions is difficult. To address this, a scatter plot can be created to focus on just two features (2D) or three features (3D). Thus, the scatter plots of 2D are used as depicted in Figures 8.36 and 8.37 to display the '*Annual_Income*' vs '*Spending_Score*' and '*Age*' vs '*Spending_Score*' respectively. The following code snippet generates a scatter plot of '*Annual_Income*' vs '*Spending_Score*' and '*Age*' vs '*Spending_Score*.'

8 • Machine learning step-by-step practical examples 195

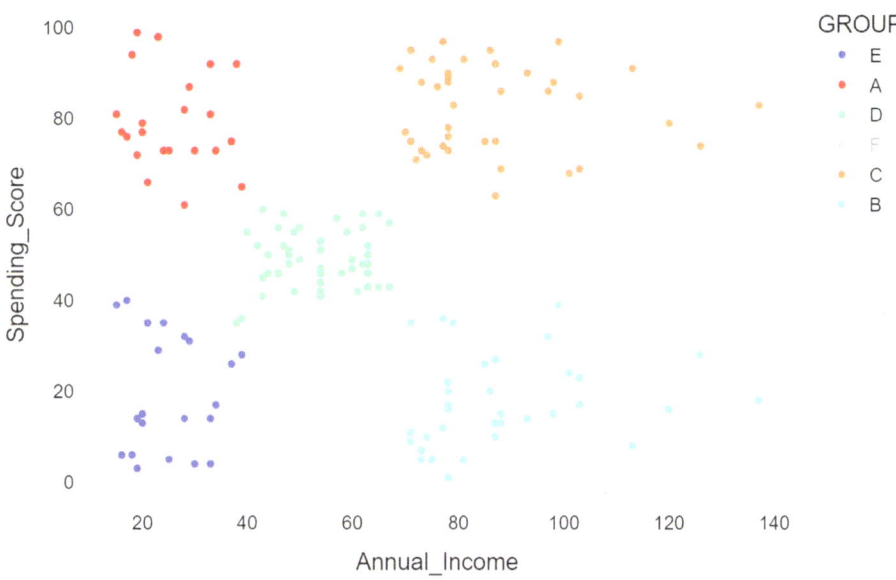

FIGURE 8.36 Scatter plot of *'Annual_Income'* vs *'Spending_Score'*.

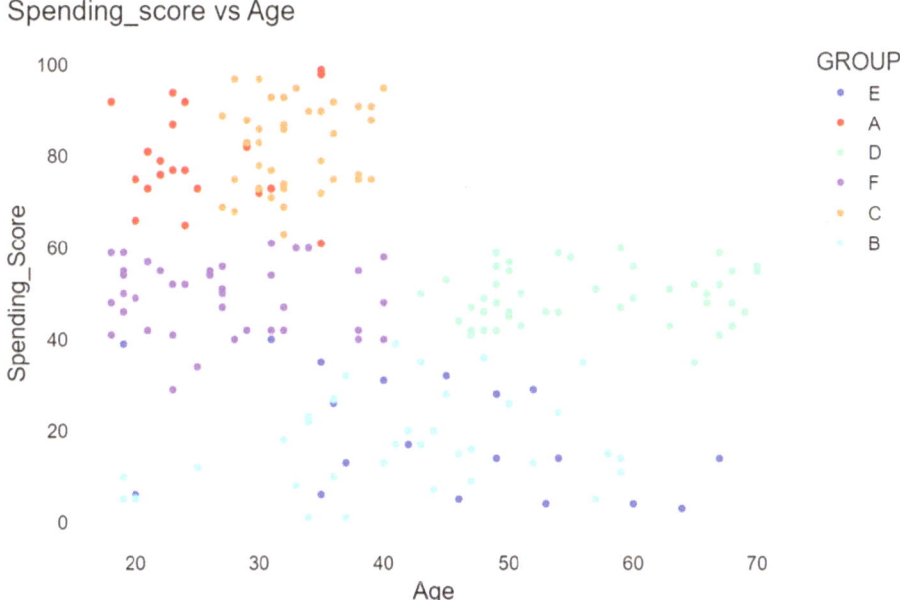

FIGURE 8.37 Scatter plot of *'Age'* vs *'Spending_Score'*.

```
fig = px.scatter(customer_data, x='Annual_Income',
y='Spending_Score',color='GROUP')
fig.update_layout(title='Annual_Income vs Spending_Score',
width=700, height=500)
fig = px.scatter(customer_data, x='Age', y='Spending_Score',
color='GROUP')
fig.update_layout(title='Spending_score vs Age', width=700,
height=500)
```

Additionally, the distribution of the six clusters can be distinctly visualized in 3D using the '*px.scatter_3d()*' command illustrated in the following code snippet, with the corresponding 3D visualization depicted in Figure 8.38.

```
fig = px.scatter_3d(customer_data, x='Annual_Income',
y='Spending_Score', z='Age',color='GROUP')
fig.update_layout(title='Annual_Income vs Spending_Score vs
Age', autosize=False,width=1000, height=800)
```

Furthermore, it is important to visualize the gender distribution in each cluster to provide the number of male and female customers in each customer segment. The following code snippet generates the gender distribution as depicted in Figure 8.39.

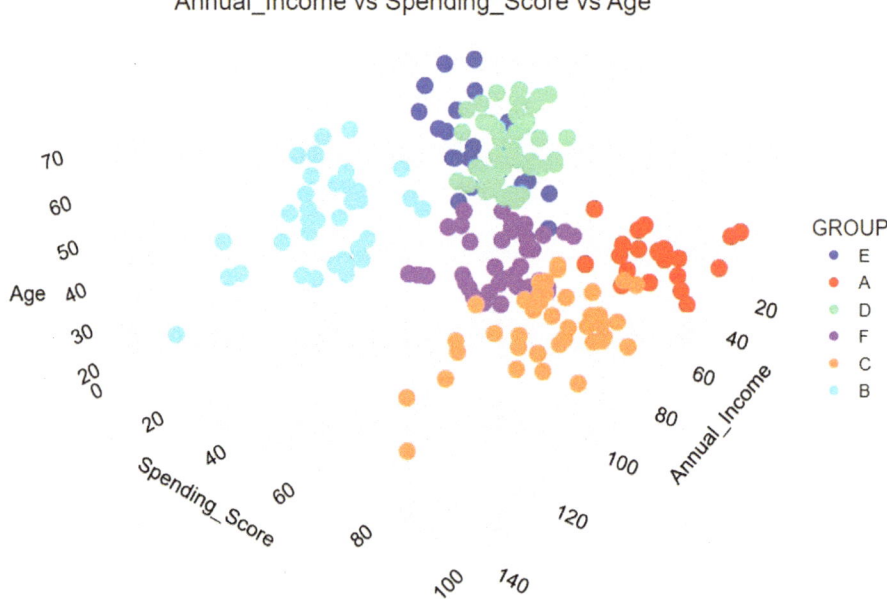

FIGURE 8.38 The 3D view of the clusters.

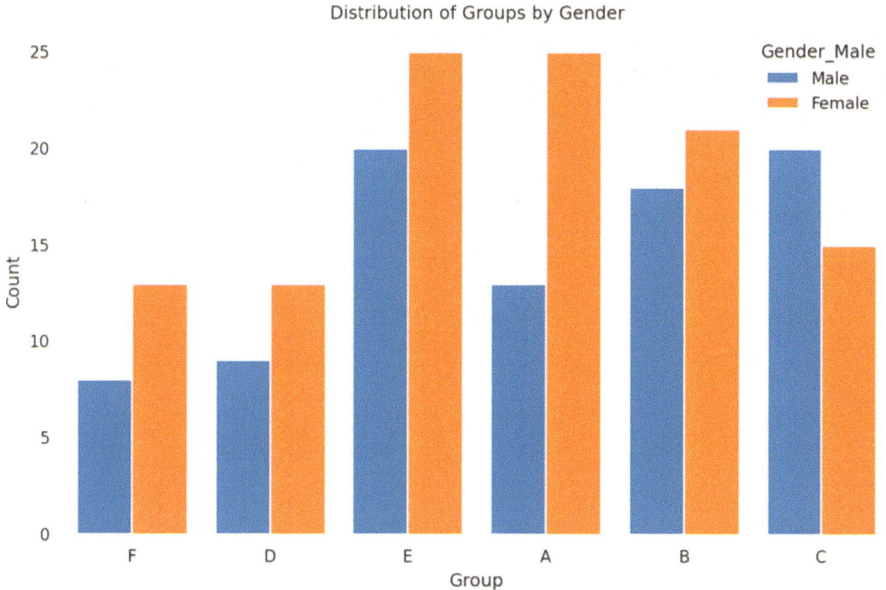

FIGURE 8.39 Distribution of gender in each cluster.

```
# Create a copy of the customer_data and replace encoded 
values with
# original ones
data_copy = customer_data.copy()
data_copy['Gender_Male'] = data_copy['Gender_Male'].replace
({0: 'Female', 1: 'Male'})
plt.figure(figsize=(10, 6))
sns.countplot(x='GROUP', hue='Gender_Male', data=data_copy)
plt.title('Distribution of Groups by Gender')
plt.xlabel('Group')
plt.ylabel('Count')
plt.show()
```

8.3.10 Model evaluation

The silhouette score using the '*silhouette_score()*' method is used to evaluate the quality of the clustering model. It measures how similar an object is to its own cluster (i.e., cohesion) compared to other clusters (i.e., separation). The silhouette score close to 1 implies well-separated clusters, near 0 indicates overlap, while close to -1 suggests misplacement of points. The following code snippet calculates the silhouette score in this case study, with the corresponding output value of 0.45206493204632353. This value suggests a moderate/reasonable separation between clusters, indicating that the data points are reasonably well-placed within their clusters but still have some degree of overlap with points in neighboring clusters.

```
from sklearn.metrics import silhoutte_score
silhouette_score_value = silhouette_score(data.drop("GROUP",
axis=1), final_model.labels_)
print("Silhouette Score:", silhouette_score_value)
```

8.3.11 Case study 4: Association rules

This case study focuses on the association rule problem, which aims to uncover meaningful insights into consumer behavior and product relationship. It illustrates the formulation of rules based on product transactions recorded within the dataset. The following subsections outline the steps in developing association rules using a given dataset.

8.3.12 Problem definition

Discovering customer purchase patterns within transactional data presents a significant challenge due to the complexity of identifying associations and relationships among items bought together frequently. Understanding the interplay of product affinities, seasonal trends, and customer preferences is crucial for optimizing product placement, enhancing cross-selling opportunities, and tailoring marketing strategies. However, the sheer volume and diversity of transactional data and the need to extract meaningful insights amid noise and variability make it difficult to uncover actionable patterns efficiently. Addressing this challenge requires sophisticated techniques such as Market Basket Analysis, which aims to identify frequent itemsets and generate association rules to guide strategic decision-making.

8.3.12.1 Description of the dataset

The dataset utilized in this case study is the Grocery Store dataset, a widely recognized and frequently employed dataset designed explicitly for association rule mining tasks. The Grocery Store dataset is a collection of customer transactions stored in a tabular format. Each row represents a single purchase, and columns include identifiers like customer ID and products. This data allows for analyzing purchase patterns by identifying frequently purchased combinations of items. It helps businesses understand customer behavior, optimize product placement, develop targeted promotions, and ultimately increase sales. The Groceries Dataset for Market Basket Analysis is publicly available for download from the Kaggle data science repository (https://www.kaggle.com/datasets/shazadudwadia/supermarket). The dataset contains 20 transactions and 11 items (i.e., '*Products*') including Jam, Maggi, Sugar, Coffee, Coke, Tea, Biscuit, Bournvita, Bread, Cornflakes, and Milk.

8.3.13 Loading libraries

The following code snippet imports the necessary libraries for this case study.

```
# Import necessary libraries
import pandas as pd
import warnings
from mlxtend.preprocessing import TransactionEncoder as TE
from mlxtend.frequent_patterns import apriori, association_rules
import matplotlib.pyplot as plt
import seaborn as sns
warnings.filterwarnings("ignore", category=DeprecationWarning)
```

8.3.14 Loading dataset

Once the essential libraries are imported, the subsequent step involves loading the dataset file (i.e., *GroceryStoreDataSet.csv*) utilizing the '*read_csv*' function within the pandas library. The following code snippet loads the dataset, and Figure 8.41 displays the first five transactions using the '*data.head(5)*' function in panda.

```
transaction_data = pd.read_csv("GroceryStoreDataSet.csv", header=None)
transaction_data.columns = ["Products"]
transaction_data.head(5)
```

Silhouette Score: 0.45205475380756527

FIGURE 8.40 Output showing the Silhouette score of the clustering model.

	Items
0	MILK,BREAD,BISCUIT
1	BREAD,MILK,BISCUIT,CORNFLAKES
2	BREAD,TEA,BOURNVITA
3	JAM,MAGGI,BREAD,MILK
4	MAGGI,TEA,BISCUIT

FIGURE 8.41 Displaying the first five products.

Number of transanctions: 20
Number of Unique Items: 11

FIGURE 8.42 Output displaying the number of transactions and unique items in the dataset.

8.3.15 Data summary

Displaying a data summary typically involves examining key statistics and characteristics of the dataset. This includes information such as the number of transactions, the total number of unique items or products available in the dataset, and the average number of items per transaction. Additionally, summary statistics might include the most frequently occurring items and measures of item popularity or support. For demonstration, the following code snippet outputs the number of transactions and unique items in the dataset.

```
# Fetch the number of transactions
num_transactions = len(transaction_data)
print(f"Number of transactions: {num_transactions}")

# Fetch the number of unique items
num_unique_items = transaction_data['Products'].str.split(',').explode().nunique()
print(f"Number of unique items: {num_unique_items}")
```

8.4 FEATURE TRANSFORMATION

Feature transformation is done to convert the transactional data into a suitable format for analysis. This is achieved by transforming the dataset into a transactional format where each row represents a unique transaction and each column represents a distinct item or product. This transformation is achieved through one-hot encoding, where the values of '1' and '0' indicate the presence and absence of an item in a transaction, respectively. Additionally, feature transformation may involve filtering out low-support items or rare items to reduce noise in the dataset and improve the efficiency of the association rule mining algorithms. Feature transformation aims to prepare the dataset for subsequent analysis and rule generation, enabling the discovery of meaningful associations between items in customer transactions. The following code snippet splits, computes one-hot encoding and displays the output shown in Figure 8.43.

	BISCUIT	BOURNVITA	BREAD	COCK	COFFEE	CORNFLAKES	JAM	MAGGI	MILK	SUGER	TEA
0	1	0	1	0	0	0	0	0	1	0	0
1	1	0	1	0	0	1	0	0	1	0	0
2	0	1	1	0	0	0	0	0	0	0	1
3	0	0	1	0	0	0	1	1	1	0	0
4	1	0	0	0	0	0	0	1	0	0	1

FIGURE 8.43 The output of one-hot encoding.

```
# Split the products in each transaction into separate items
transactions = transaction_data['Products'].str.split(',')
encoder = TE()
encoded_transactions = encoder.fit_transform(transactions)
encoded_data = pd.DataFrame(encoded_transactions.astype(int),
columns=encoder.columns_)
encoded_data.head()
```

8.4.1 Data visualization

Data visualization of unique items within the dataset typically involves creating bar charts or histograms to display the frequency of each item occurrence in the transactions. It provides a clear overview of the most commonly purchased items and their relative popularity among customers. The following code snippet computes the number of unique items in the dataset that occurred in the transaction and displays the resulting output in Figure 8.44.

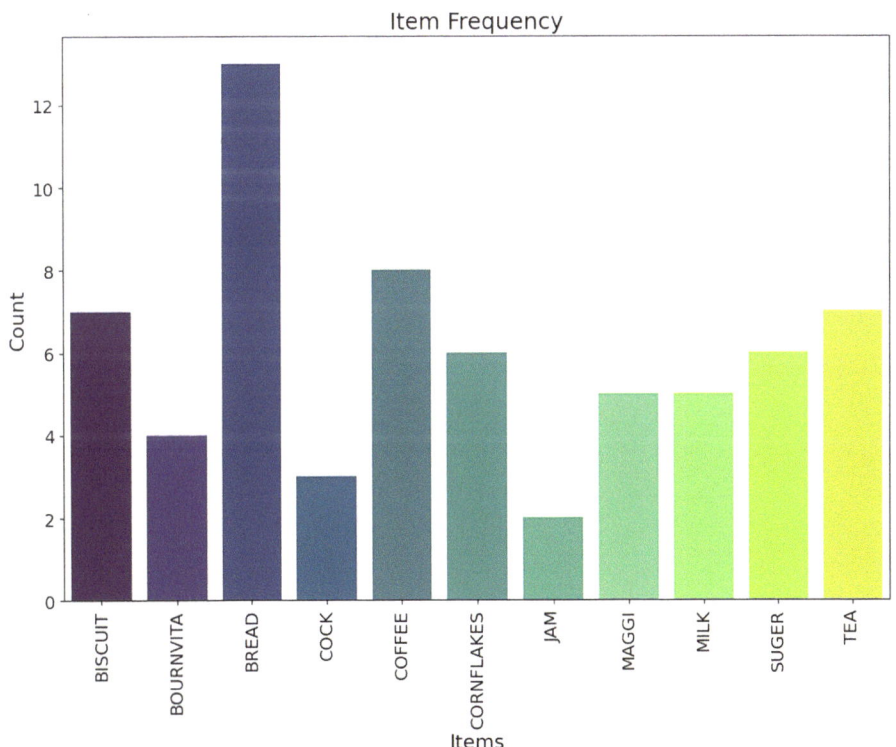

FIGURE 8.44 Frequency of items in the transactions.

```
# Bar plot of the product counts
product_counts = encoded_data.sum()
plt.figure(figsize=(12, 8))
sns.barplot(x=product_counts.index, y=product_counts.values,
palette='viridis')
plt.title('Product Counts')
plt.xlabel('Products')
plt.ylabel('Count')
plt.xticks(rotation=90)
plt.show()
```

8.4.2 Model development

In this case study, the Apriori algorithm is utilized to uncover frequent itemsets within the transactional datasets. It operates by iteratively generating candidate itemsets and pruning those that fall below a predetermined minimum support threshold. Before rule generation, the following code snippet produces combinations of itemsets ranging from single items to the maximum number appearing in transactions, as shown in Figure 8.45. Note that, for the sake of simplicity, several combinations of itemsets are omitted.

```
frequent_itemsets = apriori(encoded_data, min_support=0.1, use_
colnames=True, verbose=1)
frequent_itemsets['length'] = frequent_itemsets['itemsets'].
apply(lambda x: len(x))
frequent_itemsets = frequent_itemsets.sort_values(by='support',
ascending=False)
```

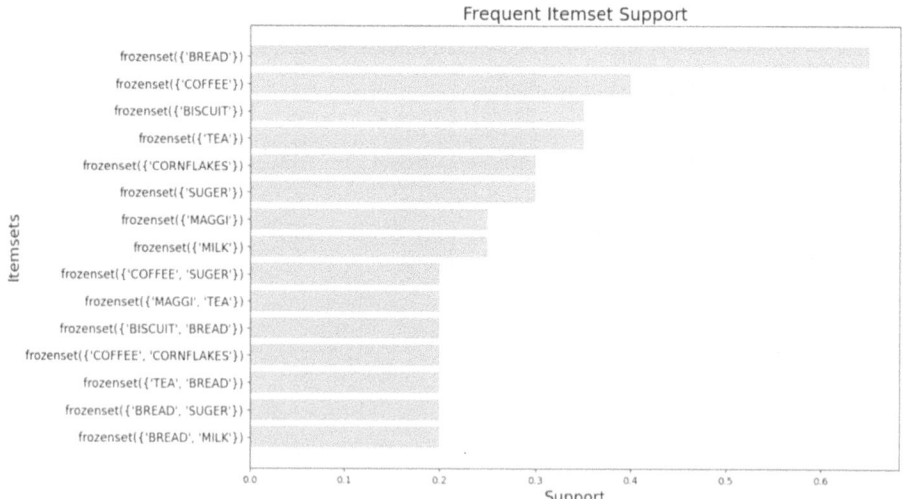

FIGURE 8.45 Combination of items in the dataset.

```
# Sort and select the top 15 itemsets. Adjust this number to
control how
# many itemsets are displayed
top_frequent_itemsets = frequent_itemsets.head(15)
plt.figure(figsize=(12, 8))
plt.barh(y=range(len(top_frequent_itemsets)), width= top_
frequent_itemsets ['support'], color='skyblue')
plt.yticks(range(len(top_frequent_itemsets)), top_frequent_
itemsets ['itemsets'])
plt.gca().invert_yaxis()   # labels read top-to-bottom
plt.xlabel('Support')
plt.ylabel('Itemsets')
plt.title('Support of Frequent Itemsets')
plt.show()
```

Once frequent itemsets are identified, association rules are generated based on these itemsets. The following code snippet computes the rules from the frequent itemset with a minimum threshold of 0.85. Then, the candidate rules are generated by combining antecedents with consequents derived from frequent itemsets. Note that there is no universally predefined minimum threshold for the '*association_rules()*' function. Setting the threshold too low can result in many meaningless frequent itemsets due to random co-occurrences, and a higher threshold will identify only the most frequent co-occurrences.

```
rules = association_rules(frequent_itemsets,
metric="confidence", min_threshold=0.85)
rules = rules[['antecedents', 'consequents', 'antecedent support',
'consequent support', 'support', 'confidence', 'lift']]
rules
```

Figure 8.46 displays the resultant association rules in a tabular format, where each row represents a rule and columns represent various metrics such as support, confidence, and lift. This allows for a concise overview of the rules and their associated metrics.

8.5 SUMMARY

This chapter explored machine learning techniques through four distinct practical case studies. It details the step-by-step practical process by employing Python programming language and a coding environment set up with Jupyter Notebook or Google Colab. In Case Study 1, the focus was on a classification problem where the objective was to detect diabetes using a machine learning classifier. This involves classifying data samples into positive or negative classes. Moving on to Case Study 2, the chapter jumped into a Regression Problem, using an advertising dataset to predict sales based on advertising budgets. This case study illustrated the relationship between advertising and sales

	antecedents	consequents	antecedent support	consequent support	support	confidence	lift
0	(COCK)	(COFFEE)	0.15	0.40	0.15	1.0	2.500000
1	(COCK, CORNFLAKES)	(COFFEE)	0.10	0.40	0.10	1.0	2.500000
2	(JAM, MAGGI)	(BREAD)	0.10	0.65	0.10	1.0	1.538462
3	(JAM, BREAD)	(MAGGI)	0.10	0.25	0.10	1.0	4.000000
4	(JAM)	(MAGGI, BREAD)	0.10	0.15	0.10	1.0	6.666667
5	(TEA, BOURNVITA)	(BREAD)	0.10	0.65	0.10	1.0	1.538462
6	(MAGGI, BISCUIT)	(TEA)	0.10	0.35	0.10	1.0	2.857143
7	(TEA, BISCUIT)	(MAGGI)	0.10	0.25	0.10	1.0	4.000000
8	(COFFEE, BISCUIT)	(CORNFLAKES)	0.10	0.30	0.10	1.0	3.333333
9	(COCK, BISCUIT)	(CORNFLAKES)	0.10	0.30	0.10	1.0	3.333333
10	(COCK, CORNFLAKES)	(BISCUIT)	0.10	0.35	0.10	1.0	2.857143
11	(COCK, BISCUIT)	(COFFEE)	0.10	0.40	0.10	1.0	2.500000
12	(COFFEE, BISCUIT)	(COCK)	0.10	0.15	0.10	1.0	6.666667
13	(BISCUIT, MILK)	(BREAD)	0.10	0.65	0.10	1.0	1.538462
14	(JAM)	(MAGGI)	0.10	0.25	0.10	1.0	4.000000
15	(JAM)	(BREAD)	0.10	0.65	0.10	1.0	1.538462
16	(COCK, CORNFLAKES, BISCUIT)	(COFFEE)	0.10	0.40	0.10	1.0	2.500000
17	(COFFEE, COCK, BISCUIT)	(CORNFLAKES)	0.10	0.30	0.10	1.0	3.333333
18	(COFFEE, COCK, CORNFLAKES)	(BISCUIT)	0.10	0.35	0.10	1.0	2.857143
19	(COFFEE, BISCUIT, CORNFLAKES)	(COCK)	0.10	0.15	0.10	1.0	6.666667
20	(COCK, BISCUIT)	(COFFEE, CORNFLAKES)	0.10	0.20	0.10	1.0	5.000000
21	(COCK, CORNFLAKES)	(COFFEE, BISCUIT)	0.10	0.10	0.10	1.0	10.000000
22	(COFFEE, BISCUIT)	(COCK, CORNFLAKES)	0.10	0.10	0.10	1.0	10.000000

FIGURE 8.46 Resultant association rules in tabular form.

to demonstrate the development of a prediction model to forecast sales outcomes. Case Study 3 shifted the focus to a clustering problem to organize unlabeled data into distinct groups based on inherent similarities or patterns. Unlike classification or regression problems, clustering algorithms categorize unlabeled data into clusters, and the chapter outlined the steps involved in developing a clustering model using the provided dataset. Finally, Case Study 4 explored association rules to uncover meaningful insights into consumer behavior and product relationships. This case study demonstrated the formulation of rules based on product transactions recorded within the dataset, providing a step-by-step guide to developing association rules and gaining insights into customer purchase patterns. Thus, through these practical examples, the chapter aimed to provide hands-on skills in applying various machine-learning techniques to real-world datasets, covering classification, regression, clustering, and association rule mining. Each case study offers valuable insights and practical guidance for understanding and implementing machine-learning models.

Exercises

1. For the classification problem in Case Study 1, analyze the dataset used for detecting diabetes and identify the key features that contribute most to the classification task.
2. In the regression problem in Case Study 2, experiment with different regression algorithms such as K-NN regression and decision tree regression, and compare their performance in predicting sales based on advertising budgets.
3. For the clustering problem in Case Study 3, apply various clustering algorithms such as agglomerative clustering and DBSCAN to the dataset and evaluate their effectiveness in organizing unlabeled data into distinct groups.
4. In the association rules problem in Case Study 4, explore different support and confidence thresholds for generating association rules and analyze how they impact the number and quality of rules discovered.
5. Implement feature engineering techniques such as feature scaling, dimensionality reduction (e.g., PCA), and feature selection on the dataset used in Case Study 1, and evaluate their effects on classification performance.
6. Experiment with different cross-validation settings on the dataset used in Case Study 1 and assess the impact on classification performance.
7. Experiment with different clustering techniques, such as Fuzzy-C-Means-Clustering and Gaussian mixture on the dataset used in Case Study 3, and compare their performance with the k-means clustering algorithm.
8. Investigate the use of association rule mining algorithms such as FP-growth and Eclat in Case Study 4, and analyze their ability to generate high-quality rules.

9. Use a publicly accessible dataset for a classification task. Experiment with a different classification algorithm, such as SVM, random forest, and naive Bayes, to perform cross-validation and compare their performance using different metrics.
10. Perform market basket analysis on a publicly available dataset similar to the one used in Case Study 4, and apply the Apriori algorithm to uncover interesting patterns of item co-occurrence in customer purchases. Propose actionable insights for improving product recommendations or marketing strategies based on the discovered rules.

Appendix
Machine Learning Resources

RESOURCE	SOURCE
Python Programming	1. Corey Schafer: https://www.youtube.com/user/schafer5 2. Sentdex: https://www.youtube.com/user/sentdex 3. Edureka: https://www.youtube.com/playlist?list=PL9ooVrP1hQOHUfd-g8GUpKl3hHOwM_9Dn 4. Python Machine Learning Tutorial: https://www.youtube.com/watch?v=7eh4d6sabA0 5. Machine Learning With Python: https://www.youtube.com/watch?v=c8W7dRPdIPE 6. Codecademy: Codecademy's Python Course is an interactive and beginner-friendly platform. It provides hands-on coding exercises to reinforce concepts. 7. SoloLearn's Python Course is a mobile-friendly platform with a community aspect, allowing you to learn and practice Python on the go. 8. *Real Python* provides tutorials, articles, and other resources that cater to developers at various skill levels. It covers both fundamentals and advanced topics. 9. The official Python website itself is an excellent resource. It provides documentation, tutorials, and links to various learning resources.

Machine Learning	1. Machine Learning with Maths, Statistics, and Linear Algebra by Andrew NG applied AI: https://www.youtube.com/watch?v=PPLop4L2eGk&list=PLLssT5z_DsK-h9vYZkQkYNWcItqhlRJLN
	2. Machine Learning by Statquest with Josh Starmer: https://www.youtube.com/user/joshstarmer
	3. Machine Learning Stanford University: https://www.youtube.com/watch?v=jGwO_UgTS7I
	4. Introduction to Machine Learning Udacity: https://www.udacity.com/course/aws-machine-learning-engineer-nanodegree--nd189
	5. Introduction to Machine Learning Yale University: https://www.cs.cmu.edu/link/research-notebook-discipline-machine-learning
	6. Introduction to Machine Learning Berkeley University:
	   a. https://ml.berkeley.edu/
	   b. https://launchpad.berkeley.edu/
	7. Google Python Class: https://developers.google.com/edu/python/
	8. Python HOWTOs, invaluable for learning idioms: https://docs.python.org/2/howto/index.html

RESOURCE	SOURCE
	9. Introduction to Artificial Intelligence (AI) by Microsoft on edX is a comprehensive program covering AI and machine learning concepts. 10. Fast.ai provides a practical and top-down approach to learning machine learning. They offer free courses that are highly regarded for their effectiveness.
Machine Learning Libraries Guides	1. Python Standard Library Reference: https://docs.python.org/2/library/index.html 2. SciPy Lecture Notes: http://www.scipy-lectures.org/ 3. NumPy User Guide: http://docs.scipy.org/doc/numpy/user/ 4. Matplotlib gallery of plot types and sample code: http://matplotlib.org/gallery.html 5. Matplotlib Beginners Guide: http://matplotlib.org/users/beginner.html 6. Matplotlib API Reference: http://matplotlib.org/api/index.html 7. Pandas documentation page (user guide). Note the table of contents on the left-hand side, it is very extensive: http://pandas.pydata.org/pandas-docs/stable/ 8. Pandas cookbook provides many short and sweet examples: http://pandas.pydata.org/pandas-docs/stable/cookbook.html 9. Pandas API Reference: http://pandas.pydata.org/pandas-docs/stable/api.html 10. The scikit-learn API Reference: http://scikit-learn.org/stable/modules/classes.html 11. The scikit-learn User Guide: http://scikit-learn.org/stable/user_guide.html 12. The scikit-learn Example Gallery: http://scikit-learn.org/stable/auto_examples/index.htm
Machine Learning Projects	Machine Learning Projects: https://www.youtube.com/watch?v=5Txi0nHle0o&list=PLZoTAELRMXVNUcr7osiU7CCm8hcaqSzGw

Machine Learning Blogs	1. Towards Data Science: https://towardsdatascience.com/
	2. Medium Machine Learning: https://medium.com/topic/machine-learning
	3. Reddit: https://www.reddit.com/
	4. Hackers News: https://news.ycombinator.com/
	5. Explainable AI:
	   a. https://www.ibm.com/watson/explainable-ai
	   b. https://www.darpa.mil/program/explainable-artificial-intelligence
	   c. https://towardsdatascience.com/explainable-ai-9a9af94931ff
	   d. https://www.weforum.org/agenda/2022/03/designing-artificial-intelligence-for-privacy/
	   e. https://ora.ox.ac.uk/objects/uuid:2b379a39-2bd9-43c1-a97a-78632ddb9ede

Appendix

RESOURCE	SOURCE
Mathematics for Machine Learning	1. MIT Courseware Linear Algebra: https://ocw.mit.edu/courses/18-06-linear-algebra-spring-2010/ 2. Calculus 3blue1brown: https://www.3blue1brown.com/topics/calculus 3. Introduction to Probability The Science of Uncertainty: https://www.edx.org/course/probability-the-science-of-uncertainty-and-data 4. Khan Academy offers a wide range of tutorials on mathematics, including algebra, calculus, linear algebra, and statistics. It provides a step-by-step approach suitable for beginners. 5. edX provides online courses from universities worldwide. Courses such as "Essential Mathematics for Artificial Intelligence" by Microsoft on edX cover relevant topics. 6. Brilliant provides interactive courses in mathematics and science. The "Mathematics for Computer Science" course is suitable for building a strong mathematical foundation. 7. Mathematics Stack Exchange is a community where you can ask questions and get answers related to mathematics. It's a valuable resource for clarifying concepts. 8. Channels like Professor Leonard and PatrickJMT offer comprehensive tutorials on various mathematical topics. 9. *Mathematics for Machine Learning* by Marc Peter Deisenroth, A Aldo Faisal, and Cheng Soon Ong is a book specifically designed for those entering the field of machine learning.
Machine Learning Algorithms	Algorithm Design and Analysis Pennsylvania University: https://repository.upenn.edu/sd3x/
Deep Learning	1. Deep Learning Andrew Ng: https://www.youtube.com/watch?v=CS4cs9xVecg&list=PLkDaE6sCZn6Ec-XTbcX1uRg2_u4xOEky0 2. CS231n - Convolutional Neural Networks for Visual Recognition is a widely praised course by Stanford University. It covers convolutional neural networks (CNNs) and their applications. 3. MIT OCW: Introduction to Deep Learning provides lecture notes and resources for learning deep learning concepts. 4. PyTorch Tutorials on the official PyTorch website provide hands-on guides for learning deep learning using PyTorch, a popular deep learning framework. 5. TensorFlow Tutorials on the official TensorFlow website offer practical guides for building deep learning models using TensorFlow. 6. *Deep Learning* (deeplearningbook.org) by Ian Goodfellow, Yoshua Bengio, and Aaron Courville is a comprehensive book that covers the theoretical foundations of deep learning.

Index

Pages in *italics* refer to figures and pages in **bold** refer to tables.

A

activation functions, 60–62, 74
    hyperbolic tangent, 61
advertising, 175–177, 180–181, 203
AGI (Artificial General Intelligence), 152–160
algorithm design, 146–147, 209
algorithms, 3–8, 13, 65, 74, **79**, 80, 94–95, 97, 104–105, 107–108, 118, **120**, 121–123, 129, **130**, 132, 143–147, 150, 158, 172
ANI (Artificial Narrow Intelligence), 152, 159
anomalies, 7, 12, 60
APIs (Application Programming Interfaces), 106, **114**, 123, 128
apriori algorithm, 202, 206
artificial general intelligence, 152–160
artificial intelligence, 6–9, *10*, 16, 145, 147–148, 150, 152, 208–209
artificial intelligence systems, 139, 143
Artificial Narrow Intelligence, *see* ANI
ASI (Artificial Super Intelligence), 152, 159–160
association rule mining, 97–98, **120**, 205
association rules, 5–6, 103, 126, 198, 203
AUC-ROC values, 100

B

Bayes' Theorem, 51
Bernoulli distribution, 54
bias and discrimination, **142**
bias term, 61–62
biases, 14, 74, 82, 92, 94, 107, **127**, 129, 138, 141–143, 145–147, 149–150, 157
big data, 94, **120**
binomial distribution, 55
box plots, 36–37, 40, 83, 87, **114**, 169, 180

C

Caffe, **122**
categorical data, 34, 40, 83, 87
Central Processing Unit (CPU), 133–134, 136
central tendency, 33–34, 52, 73, 83, 87, 165, 170, 177, 188
Centroid for Cluster, 70–71
centroids, 65, 67–71

chatbots, 2, 8–9, 11–13, 134, **142**
Chebyshev distance, 65, 67
class distribution, 166–167
classes, 3–4, 34, 63, 83, 92, 99–101, 162, 165, 174
classification, 3, 97–98, 113, **114**, **116**, 118, **120**, 121, 123, 126, 162, 184, 205
classification algorithms, 3, 162, 172, 206
cloud computing, 131, 135
cloud computing services, 132, 135
clustering, 5, 65, 83, 97–98, 105, 113, **114**, **116**, 118, **120**, 121–123, 126, 184, 205
clustering algorithms, 5, 65, 162, 184, 205
clusters, 5, 60, 65, 67–71, 102–103, 184, 192–194, 196–198, 205
code, **79**, 105, 112–113, **122**, 125, 128, 164, 186–187
code editors, 112–113, **116**, 117, **119**, 121
code snippet, 164, 169, 176, 180, 183, 193
computer science, 7, 153, 156
computer vision, 7, 9, *10*, **79**, **122**, 158
conditional probability, 50–51
conditional probability of event, 50–51
confusion matrix, 98, 172–173
continuous random variables, 51–54, 56
convolutional, 209
correlations, 12, 140, 167, 175, 177, 190
cost, 59, 82, 110, 113, 135–136

D

data, 1, 7–8, 33–36, 38–41, 74, 76–88, 91–97, 106–107, 139–143, 148–149, 166–167, 169, 171–172, 174–176, 180, 182, 186–188, 192–193, 196–202
data collection, 76, 78, 81–82, 92–94, 140–141, 150
data curation, 80, 93
data distribution, 40, 62, 88
data labeling, 80, 131
data mining, 12, 25, 94
data points, 5–6, 19, 35–39, 41, 52, 60, 65, 67–71, 83, 102–104, 177, 184–185
data preparation, 76–77, **79**, 80–94, 118
data preprocessing, 76, 81, 88, 118, **122**, 126, 169, 171, 180
data repositories, **79**, 93–94

Index 211

data samples, 3–4, 39, 71, 105, 141, 162, 194
data science, 6–7, 13, 16, 19, 25, 53, 94, 208
data scientists, 14, **79**, 111
data sources, 78, 93, 143–144
data standardization, 171
data summary, 165, 176–177, 186–187, 200
data transformation, 83, 93, 163, 180, 182
data visualization, 87, 126, 166, 177, 187, 201
dataset, 5, 32–33, 35–41, 78–85, 87–88, 91–99, 101–102, 104–105, 128–129, **130**, 141, 163–167, 169–172, 175–177, *178*, 180–182, 185–188, 198–202, 205
dataset features, 169, 181
dataset file, 164, 176, 185, 199
dataset splitting, 95
DBSCAN (Density-Based Spatial Clustering of Applications with Noise), 5, 205
decision trees, 4, 172
deep learning, 8–9, 11, 16, 151, 154, 209
deep learning models, **114**, 123, **142**, 209
deployment, 106–110, 112, **114**, **122**, 125, 128–129, 131, 138, 143, 146, 148–149, 159
discrete random variables, 51–54, 63

E

EDA (Exploratory Data Analysis), 87–88, 93, **116**
eigenvalues, 18, 30–32, 74
eigenvectors, 18, 25, 30–32, 74
EOA (Evolutionary Optimization Algorithms), 104–106, 108–109
error term, 45, 47–48
ethical challenges, 146, 155
ethical frameworks, 107, 143, 150
ethical issues, 82, 108
Euclidean distance, 60–61, 65, 68
Euclidean space, 65–67
evaluation metrics, 98, 108
events, 48–51, 55, 124, 140

F

False Negative (FN), 98–99, 173
false positive, 98–99, 173
feature extraction, 92, 115
feature selection, 63, 91, 93–94, 104, 115, 181, 205
forecasting, 12, 101
functions, 7, 53, 58–59, **86**, 118, 143, 154, 158, 174, 176, 185, 199, 203

G

GA (Genetic Algorithms), 106, 121
Gaussian distribution, 40, 56
Generative Pre-trained Transformer (GPT), 157
Genetic Algorithms (GA), 106, 121
GPUs (Graphics Processing Unit), 133–136

gradient, 59–60, 105
gradient descent, 59, 104–105

H

hyperparameter optimization, **130**, 137
hyperparameter tuning, 115, **116**, **120**, **130**

I

IDEs (Integrated Development Environments), 110–113, 117–**119**, 121, **122**, 133, 136
income, 186, 188, 194–196
independent variables, 4, 45–47, 59, 102, 181, 183
inferential statistics, 33, 39, 43
information theory, 62
input features, 3–4, 91, 171, 177, *179*, 181
input variables, 59
intelligence, 7, 135, 152–156, 158–159
intelligent systems, 121

J

JAX (Just Another X), 134
joblib, 174
Jupyter Notebook, 112–113, 135, 203

K

k-means, 5, 194
k-means algorithm, 192–193
Keras, **114**, **127**, **130**
kNN, **114**, **116**, 132, 172
knowledge, 1, 3, 12–13, 124, 152–153, 159

L

labeled data, 4, 6, 8
large language models (LLMs), 157, 159
learning algorithm, 3, 105
libraries, 110–113, **114**, 115, **116**, 118, **120**, 121–125, **127**, **130**, 136–137, 163, 174–176, 208
linear algebra, 13, 18, **114**, 123, 207, 209
LinearRegression, 182
loading libraries, 163, 175, 185, 199
loss function, 64, 105

M

machine learning, 1–33, 39–41, 49–51, 53–65, 73–74, 92–94, 108–109, 121–124, 126, 134–137, 139–**142**, 149–150, 207, 209
machine learning algorithms, 3, 12, 18–19, 40, 72, **79**, 80–81, 83–84, 87, 108, **114**, **116**, 123, 141–**142**
MAD (Mean Absolute Deviation), 37–39

Index

MAE (Mean Absolute Error), 98, 101, 183
Manhattan distance, 65–66, 69
mathematical function, 51, 53, 60, 97
MATLAB, 110–111, 115, 117, 123
matrices, 18, 20–22, 25–26, **114**
matrix, 20–27, 29–30, 98, 115, 164, 173, 190
Mean Absolute Error, *see* MAE
Mean Percentage Error (MPE), 183
mean squared error, *see* MSE
mean values, 42–43, 102, 193
measures of dispersion, 34, 38, 73
median, 34, 36, 39, 83–84, 88, 167, 170–171, 188
MLOps, 107–108
model, 1–4, 6, 33, 58–60, 76, 80–81, 91–92, 95–102, 104–106, 108–109, 126, **127**, 129, **130**, 132–134, 139–143, 148–150, 157, 171–175, 181–185, 193
model deployment, 106, 108, **127**
model equation, 183
modeling phases, 77, 180
MPE (Mean Percentage Error), 183
MSE (Mean Squared Error), 98, 101–102, 183–184
multivariate analysis, 89, 91
multivariate linear regression, 47, 182

N

natural language processing, 7–9, *10*, 115, **120**, 121, 152
neural networks, 8, 61, 64, 97, **114**, **120**–121, **122**, **127**, 154, 156
NLP (Natural language processing), 7–9, *10*, 16, **79**, 115, **120**, 121, 123, 152
noise, 5, 81, 141, 186, 198, 200
normal distribution, 40–43, 56–57
normalization, 33, 50, 84–85, 93, 182

O

object recognition, **114**
one-hot encoding, 84, 191, 200
open-source version control system, 129, 133
optimization, 104, **114**, **120**, 126, 129–131, 151, 154, 156
optimization algorithms, 104, 113
outcomes, 48–51, 53–54, 56–57, 59–60, 63, 77, 98–99, **142**, 143, 158–159, 163, 166–167, 171, 174
outliers, 5, 33–34, 36, 39, 81, 83, 85, 88, 166–169, 174, 180

P

parameters, 48, 65, 104–106, 140–141
PCA (principal component analysis), 74, 91–94, 205

PDF (probability density functions), 53–54, 57
performance metrics, 98–102, 126, 183
Poisson distributions, 54–55
precision, 82, 98–100, 172–173
predict, 3–4, 47–48, 72, 97, 102, 104, 173–175, 183–184
pre-trained models, **79**, **114**, **120**, **122**, 126, 128, 136
principal component analysis, *see* PCA
privacy, 14, 82, 107–108, 139–140, **142**, 143, 146–147, 149–151
probability, 13, 48–51, 53–56, 72, 172, 209
probability and statistics, **13**
probability density function, 52–54, 57
probability distribution, 49, 51, 53–54, 57, 62–64
probability mass function, 52–55
probability theory, 48–49, 51, 53
problem domains, 12, 77–78
programming, 113, **116**, 117, 121, 124, 153, 207
programming code editors, 113, 121, **122**
programming languages, 110–113, 117, 121, 123–126, 136
programming libraries, 115, **116**, 121–123
Python, 110–113, 117, 123, 131, 163, 165, 207
Python code editors and IDEs, 111–113
Python libraries, 113–115
Python programming, 113, 131
Python tools, 111
PyTorch, 110, **114**, **127**, 134, 209

Q

quartiles, 36–37, 73, 167

R

R-Square, 183
RAM (Random Access Memory), 133–134, 136
Random Access Memory, *see* RAM
random variables, 49, 51–57, 62–63
recommendation systems, 2, 12, 61, 77–78, 123
regression, 3–4, 33, 56, 97–98, 113, **114**, **116**, 118, **120**, 121–123, 126, 175, 205
regression algorithms, 4, 162, 180, 182, 205
regression models, 101, **116**, 175
reinforcement learning, 3, 6, 158
RMSE (Root Mean Squared Error), 98, 102, 183
ROC curve, 100
Root Mean Squared Error, *see* RMSE
rows, 19–22, 24, 27, 87, 165, 175–176, 198, 200, 203

S

scalability, 118, **122**, 123, 125, 134, 136, 154, 156
scatter plots, 87, **114**, 167, 169, 177, *179*–180, 190, 194, *195*

semi-structured data, 78
semi-supervised learning, 3, 6
sensitivity, 98–101, 172
SGD (Stochastic Gradient Descent), 104–105
sigmoid function, 60
silhouette score, 102–103, 108, 197–199
slope, 45–47, 59–60, 105, 177
SMOTE (Synthetic Minority Oversampling Technique), 93
software tools, 110, 132, 136
standard deviation, 38–42, 52–53, 57, 85, 87, 174
statistics, 13, 32–33, 40, 50–51, 53, 78, **127**, 165, 207, 209
Stochastic Gradient Descent, *see* SGD
structured data, 78, 175
supervised learning, 3, 6, 80
support vector machine, 4, 121, 132, 172

T

TensorFlow, 110, **114**, **120**, **127**, 134, 209
testing data, 106
testing sets, 95, 97–98, 104, 108, 171
tokenization, **120**
TPUs (Tensor Processing Unit), 134–136
training, 6, 95–97, 104–106, 108, **114**, 123, 128, 134–135, 153–154, 171–172, 174, 182
training and testing data, 106
training and testing sets, 95, 97, 104, 108, 171
training parameters, 104, **127**
training set, 95–97, 104, 171–172, 182
transactions, 103–104, 198–202
true negative, 98–99, 173
true positive (TP), 98–99, 173

U

uniform distribution, 56–57
unlabeled data, 6, 65, 184, 205
unseen data, 3, 96, 98, 104, 108, 174
unstructured data, 8, 78, 154
unsupervised learning, 3–5, **122**

V

vectors, 18–21, 59
visualization, 39, **79**, 87, 112, **114**, 117, 126, **127**, 186, 193, 196
visualization techniques, 33, 83
visualization tools, **122**

W

weights, 39, 61–62, 74, **127**, 129, 163

Z

zscore, 171, 174